The Scientific Conquest
of Death

IMMINST.ORG

The Scientific Conquest
of Death

Essays on Infinite Lifespans

Immortality Institute

www.librosenred.com

CDD 814   Immortality Institute
           The scientific conquest of death : essays
           on infinite lifespans. – 1a. ed. – Buenos Aires :
           LibrosEnRed,2004.
           296 p. ; 22x14cm.

           ISBN 987-561-135-2

           1. Ensayo Estadounidense I. Título

C.E.O.: Marcelo Perazolo
Contents Director: Ivana Basset
Interior Design: Vanesa L. Rivera

Commissioning Editor: Bruce J. Klein
Editor in Chief: Sebastian Sethe
Editorial contributions: Devon B. Fowler, Justin Loew, Reason,
Casey S. Tompkins
Cover Art: Tyrone Pow

This book was initiated, compiled, and edited by members of the online organization, Immortality Institute (http://imminst.org).

Design, typesetting, and other prepress work by LibrosEnRed
www.librosenred.com

© LibrosEnRed, 2004

ISBN: 987-561-135-2

First english edition - Print on Demand

TABLE OF CONTENTS

INTRODUCTION

The mission of the Immortality Institute is to conquer the blight of involuntary death.

Some would consider this goal as scientifically impossible. Some would regard it as hubris. Others say: "Don't mention the 'D–word', it will just scare people, and turn them away from the very real possibility that modern science will help us to dramatically extend our healthy life span."

What should we make of this? Is it possible that scientists – or at least humankind – will "conquer the blight of involuntary death?" If so, to what extent will we succeed? What is in fact possible today, and what do the experts predict for the future? Is such a thing as 'immortality' feasible? Moreover, is it desirable? What would it mean from a political, social, ethical and religious perspective?

This book will help to explore these questions.

When the Institute was approached regarding the possibility of distilling some of the lively and insightful debates that take place within its online forums into book form, questions arose over what such a book should contain. In the last few years, a couple of very good books on the scientific conquest of death have been published. (These are indexed in the bibliography at the end of this work.) How would this book be special?

After careful consideration, the answer seemed clear: This should be the first truly *multidisciplinary* approach to the topic. We would discuss not only biological theories of aging, but also biomedical strategies to counter it. Moreover, we would consider alternative approaches such as medical nano-technology, digitalization of personhood, and cryobiological preservation. But this would only be part of the whole. We also wanted to tackle some of the questions that are usually left unanswered in the last chapter of scientific books: If we accept that radical life extension is a real scientific possibility, then where does that leave us? Would it create overpopulation, stagnation and perpetual boredom? How would it change our society, our culture, our values and our spirituality? If science allows us to vastly extend our life span, should we do so?

It became clear that a single author, however knowledge-able, could not possibly address this kaleidoscope of topics adequately. Thus, we decided to publish a compilation of essays. Some stem from an open call for papers, some are invited contributions by established authorities in a particular field, and a few are specially selected reprints. From among the numerous contributions, we carefully choose the best in our eyes. Considering the multitude of topics and the quality of the submissions, it was an exceptionally difficult task. The result can only ever be a compromise. A compromise between conveying scientific information adequately, and accessibility to the lay reader; between philosophical depth, and the desire to stress relevancy; and, of course, between limitless curiosity, and the very limiting constraints of space. We hope that you like the result.

OVERVIEW

This book is divided into two sections: science, (including biology, biomedicine, nanotechnology, digitalization and cryonics) and perspectives (including literature, history, philosophy, sociology and ethics). This is not a strict division, as scientific possibilities are the starting point for all philosophy, and, in turn, the scientists in this book are not blind to the philosophical implications of their work.

All essays are followed by their relevant citations. All web hyperlinks are valid as of April 2004. Please do not hesitate to call the Institute if a link is out of date, as we might be able to help chase it down. Please also note that the Institute provides additional graphics, charts, and other relevant material online and free of charge to all purchasers of this book at http://imminst.org/book1.

This book concludes with remarks, an extensive bibliography for further reading, information on the contributing authors, and a few words of thanks.

But – as we shall soon learn – there is no time to waste: Follow us into an exploration of the scientific conquest of death.

The road to immortality is just the turn of a page away.

CHAPTER I: SCIENCE

Biomedicine, nanotechnology and other strategies

We start, as is proper, by defining the subject matter. What is immortality? How can we define it in a scientifically sensible manner? Is immortality even biologically possible? These and other questions will be addressed in **"Biological Immortality"** by **Michael R. Rose**, Professor of evolutionary biology at the University of California, Irvine and author of *Evolutionary Biology of Aging*, a book that created a complete departure from the views that had dominated the field of aging research since the 1960's. We will learn that far from being a scientific impossibility, there are now good reasons for thinking that biological immortality is fundamentally possible.

If aging can, in theory, be conquered – how would, how should we go about it? **Aubrey de Grey**, an authority in the field of anti-aging theory from Cambridge University, outlines a general strategy for proceeding with "**The War on Aging.**" In his essay, Aubrey de Grey touches on numerous issues, both scientific and societal, that will be taken up later in the book.

After these introductions, we move to consider individual aspects of this strategy. Firstly, on the biomedical side, microbiologist **João Pedro de Magalhães** provides a summary overview of how "**The Dream of Elixir Vitae**" might be realized.

One of the most topical and promising approaches to extending healthy life span is stem cell tissue engineering.

Michael West, president of Advanced Cell Technology and one of the "founding fathers" of modern stem cell research has written **"Therapeutic Cloning."** It gives us an exciting insight not only into the scientific background, but also into his very personal experiences and hopes in relation to the conquest of death.

While stem cell research is still an immensely dynamic field, we have recently seen the emergence of another exciting area of potential anti-aging treatments: **"Nanomedicine"** – the science of creating medical devices through nanoscale and eventually molecular manufacturing – has received intense media scrutiny and generous public funding in the US and Europe. **Robert A. Freitas**, a true pioneer in this area, describes how tiny machines could be effective in the conquest of death. As a special bonus, a second part of this chapter, including numerous illustrations, is published online - http://imminst.org/book1.

Once inspired to consider molecular-sized machines, we are not limited to healing and repairing our aging bodies: **Ray Kurzweil**, well known futurist and the recipient of the 1999 US National Medal of Technology introduces us to **"Human Body Version 2.0,"** where advanced technology constructs and defines the very substance that we are made of.

This introduces a second vision of immortality, one that ventures beyond biology. **Dr. William Sims Bainbridge**, Deputy Director for the Division of Information and Intelligent Systems at the National Science Foundation, considers how digital information about personality, feelings, likes and dislikes can be recorded. By archiving the uniqueness of a person, we might achieve some **"Progress toward Cyberimmortality."** But can we be more ambitious? Will we one day be able to copy our 'selves' onto a computer?

"Will Robots Inherit the Earth?" asks **Professor Marvin Minsky**, who in 1959 co-founded what became the MIT

Artificial Intelligence Laboratory. He proposes humankind will indeed leave behind the constraints of biological mortality – not only conquering death, but also expanding in consciousness.

"All very well," one might contend. But will these predictions come true within our own life spans? For those who seek to conquer death, dreams of a distant future might not suffice. However, **Dr. Brian Wowk**, physicist and cryobiologist introduces us to "**Medical Time Travel**" via cryopreservation – the science of maintaining the human brain until the scientific predictions that we were considering in this section have indeed come to pass.

Biological Immortality

Michael R. Rose, Ph.D.

There are no superstitions like the superstitions of the professional biologist. Biology is a sprawling field no single person can master in its entirety. Ask a professor of biology a question about an area where they have never performed research, and I will guarantee you that their answer will contain a fair admixture of incorrectly recalled material from a long defunct textbook, wild guesses, and pure superstition. Some Professors might remember reviewing that subject for an oral exam held midway through their doctoral program, but it was so long ago that they would feel more comfortable if you let them consult a textbook from their straining shelves. The decay of their knowledge, still more their ignorance of recent research developments, is palpable. There is little likelihood that they will confidently pronounce facts of dubious value.

Not so where aging and immortality are concerned. Biologists are generally of the opinion that aging is a deep mystery, which no reasonable faculty member should research.

> "All things age, so it is probably something to do with the laws of thermodynamics, or perhaps protein breakdown. Wasn't there some problem with aging in cells that that fellow Hayburn, Hayflick, Haysomething

17

> showed? Can't be stopped, definitely inevitable, so no point worrying about it."

And with further questioning about immortality, they will reply as follows.

> "Immortality, you ask? Well, that's a Greek myth, isn't it? Nothing to do with biology. Everything dies, so immortality can't be a real thing, no way. It's a joke played on us by religions."

So saying, Professor Corngold will shake his head with relief, grateful that there was at least one question he could answer without having to refer to a textbook.

AN APPROPRIATE TERMINOLOGY FOR AGING AND IMMORTALITY

Yet, very few of the superstitions that biologists and medical doctors believe about aging, immortality, and death are true. To explain this, some refinements of terminology are required:

The most objective definition of aging is that which occurs when rates of survival or reproduction inexorably decline, even when organisms are kept in excellent environments, in which contagious disease has been virtually eliminated, with abundant food and no prospect of being eaten. [1] Some people like definitions of aging that include falling performance, increasing pigment levels, reduced cell replication, and so on indefinitely. But these definitions depend on the particulars of the biology of the organism. By contrast, all organisms have rates of survival and reproduction and when survival probabilities fall to low enough values, organisms die. It is also important to exclude all reasonable external agents of death. Needless to say, human mortality patterns are not necessarily a reflection of aging, on this definition. Even in Western populations during the 20th Century, there are dramatic spikes in mortality associated with the 1918 outbreak of 'Spanish' influenza, World War I and World War II. [1] Surely such mortality is not to be confused with the mortality associated with aging? No professional biologist would imagine that there will ever be organisms that are immune to all possible causes of death or sterility. The surface of the sun is going to kill all terrestrial life that happens to find itself there without protective equipment. If there is such a thing as biological immortality, it cannot mean survival under all conceivable conditions.

Instead, we can describe immortality more sensibly as a feature of rates of survival or reproduction. An important side issue is whether or not fertility should be included with survival in definitions of aging. For some medical professionals, the loss of fertility with age in both men and women is a clear manifestation of aging. For other professionals, it is merely incidental.

If we use a definition of aging based on declining survival and fertility, we can define immortality intelligibly. If aging can be defined as the persistent decline of these biological variables, then it makes sense to define immortality as a property of organisms that do not exhibit such declines. They may have never exhibited declining survival and reproduction, or they may have reached a point of equilibration at which further sustained declines have ceased.

Having defined aging and immortality in a concrete way, we can now proceed to discuss them in a roughly empirical manner.

SOME FACTS ABOUT AGING AND IMMORTALITY

Let us proceed to demolish superstitions. Is aging universal?

Clearly not. If everything aged, the continued survival of the cells responsible for producing our sperm and eggs (the 'germ-line') over millions of years would have been impossible. Most of the bananas that you have eaten in your lifetime come from immortal clones produced on plantations. Even in organisms like mammals, which have germ-lines that separate very early from the rest of the body, the survival and replication of the cells responsible for producing gametes (germ cells) have proceeded for hundreds of millions of years. Life can continue indefinitely.

But even if life can propagate itself indefinitely, are there any organisms that are free of aging, living with biological immortality? I must be clear about one point concerning death: It is not true that aging is required to kill organisms kept in the laboratory. Showing that a species dies in the lab is not the same thing as showing that immortality does not occur in that species. Mechanical accidents in the lab will kill many soft-bodied plants, animals, and microscopic creatures. Deadly mutations can kill at any age or time. It is also impossible to keep living things free of all diseases indefinitely. Being free of aging does not imply the complete absence of death. Biological 'immortals' will often die, just not because of a systematic, endogenous, ineluctable process of self-destruction. Death is not aging. Biological immortality is not freedom from death.

Instead, the demonstration of immortality requires the finding that rates of survival and reproduction do not show aging. There are many cases where such patterns are inferred anecdotally among plants and simple animals, like sea anemones. But the best quantitative data known to me were gathered by Martinez [2], who studied mortality rates in Hydra, the aquatic animal that used to be a staple of high school biology. Martinez found that his Hydra showed no substantial fall in survival rates over very long periods. They still died, but not in a pattern that suggested aging. Other scientists have gathered comparable data with small animals. [e.g. 3] Some species were immortal and some were not. The immortal species reproduced without sex.

Also, invoking laws of thermodynamics as limits to life is clearly incorrect, given the evolutionary immortality of life forms. Such an invocation was always rankly amateur in any case, since these laws only apply to closed systems. Life on Earth is not a closed system. The earth receives an abundant input of energy from the sun.

Thus, some of the deepest prejudices of professional biologists concerning immortality are certainly false. Aging is not universal. There are organisms that are biologically immortal. However, this is not a very substantial conclusion. We knew enough about the biology of death decades ago to come to this finding. In a sense, what I have just described is the 'old immortality', the immortality that has always been with us. Below, I will describe a new immortality that has come into view only in the last decade or so.

RECENT RESEARCH ON THE DEMOGRAPHY OF IMMORTALITY

Another superstition of most biologists is that, once aging starts, it continues implacably until every organism is dead. But is aging implacable?

In terms of mortality rates, we can re-state this question as whether or not aging species are subject to continually increasing mortality rates throughout adulthood. A deceleration in mortality was found first in human data. [4;5] Very old people did not die off as fast as demographers expected. But human survival data are hard to interpret scientifically, given the confusing effects of wars and improvements in medical practice. It might be that the considerable benefits of medical practice have been especially profound among the oldest old. Human cohorts do not supply good data for scientific inference. Human data were never going to give a clear answer to the question of the implacability of aging.

Things changed dramatically in the early 1990's when insect cohorts were used to study mortality rates late in life. [6;7] In caged insects, kept under good conditions, mortality rates stop increasing in late life. [8–10] The new facts of death reveal three phases of mortality: juvenile, aging, and late life. In the juvenile period, mortality rates do not show sustained increases. In the aging phase, mortality rates increase rapidly. In the third phase of life, mortality rates are roughly constant, though they tend to maintain a very high level. Organisms that reach the third phase can be said to be biological immortal, in that they no longer age. This is the new immortality that I wish to introduce to discussions of the attainment of human immortality.

But before doing so, I want to put the new immortality into a general biological context. One way in which we can classify organisms is to divide those that are always immortal from those that have a period of rapidly increasing mortality rates – aging –prior to a period of immortality. The aging species age first and then are immortal. Immortality is thus the universal condition of life. Aging is the less common condition. This is a fact of enormous significance for the long-term future of the human species.

What is Going On?

The evolutionary theory of life-history, aging, etc is not complete, but it is the best theoretical foundation for understanding the phenomena of immortality, old and new. Natural selection is an amazing thing, but it is not all-powerful. When natural selection is weak, survival and reproduction should be imperiled; and natural selection is relatively weak later in life. Consider genes that kill everyone who has even one copy of the killer gene. If the gene kills during childhood, it will be eliminated from the population in a single generation. Under these circumstances, natural selection is all-powerful and this is a common intuitive understanding of the action of natural selection. But if killer genes act during old age, they will not be resisted by natural selection. The killer genes have already made it into the next generation. The future survival of the victim of the killer gene does not matter for the transmission of the killer genes. Natural selection is too feeble at advanced ages. [10;11]

At early ages, natural selection is strong, and survival rates will be high, but not 100 percent, because we do not need aging to have death. During reproductive adulthood, natural selection becomes progressively weaker. This increases mortality rates, and thus causes aging, even under ideal conditions. These ideas explain both the health of youth and the advance of aging during adulthood.

But there is a subtle twist that explains eventual immortality. The force of natural selection steadily falls with adult age. But it cannot fall forever, because it cannot take on negative values. As a result, it finally hits zero, and stops falling, near the end of life. The force of natural selection reaches a plateau from which it does not budge. This explains why immortality arises in late life. As natural selection cannot get any worse, bad evolutionary effects stabilize. Since these bad effects are the cause of aging, their stabilization stops aging. These are just words, but explicit mathematical theory shows that the arithmetic of natural selection can produce an immortal period late in life. [13–15]

Evolutionary theory can explain immortality in organisms like flies and ourselves. It also explains why some organisms are always immortal. Evolutionary theory predicts that organisms that reproduce by splitting their bodies into two similar parts will not age because there is no weakening of selection. After the single reproductive act, two juveniles are recovered, with no adult to undergo aging. Natural selection stays in its first phase, at all ages. These species always have biological immortality.

The Manipulation of Aging and Immortality

Evolutionary biologists have been deliberately postponing aging since the 1970's. This has been accomplished by delaying the period of weak natural selection over many generations by delaying reproduction. Delayed reproduction prolongs the phase when natural selection is strong. Evolution then does all the work once the pattern of natural selection has been changed and aging is readily postponed. [1, pg.39–]

But there is a completely different approach to the manipulation of life span. What if the mortality plateau could be brought forward to earlier ages, at earlier levels of physiological robustness? Individuals that have been medically treated in this manner would no longer age. The rate at which they would die would depend on the age at which the transition to immortality occurs. If the transition occurs at an early age, this rate might be low. That is, immortality would be fairly healthy. This idea was published some time ago, in the 1970 science fiction novel *One Million Tomorrows* by

Bob Shaw. [16] Shaw assumed that the cessation of aging would require the termination of sexual function. But we have no reason to accept such a connection between reproduction and immortality at this time. When Shaw wrote his novel, the idea of human immortality was literally science fiction. But in my laboratory we now have data showing this type of transition in experimental organisms – fruit flies – using selective breeding. [17] We have not fully eliminated aging, but we have both decelerated aging and reduced the level of mortality during the immortal phase of life. Toward the end of life, some of our flies even cease aging earlier.

The immortal phase can be manipulated, just as aging can be. In particular, immortality can be made to come earlier. This does not mean that the methods used to produce this effect in the lab can be applied to humans, but the results show what is biologically possible in simple animals.

THE POSSIBLE

What is the practical situation with respect to attaining human immortality earlier, in better health?

Later life is not adaptive. It is instead a product of evolutionary accidents. Such accidents involve many bits and pieces of biology, because they are not focused by selection. To attain an earlier, more beneficent immortality requires the re-shaping of multiple features of our biology, not a few. Unlike Bob Shaw's story, in real life, a single drug or treatment will not be enough to give us immortality when we are still healthy enough to enjoy it.

Fortunately, 21st Century biology can supply us with thousands of pharmaceuticals and other medical treatments, especially when the full power of genomics is applied to the development of drugs. While the simple magic elixirs of horror movies will never successfully control aging and immortality, we do not need to find such potions. Modern automobiles have thousands of parts, precisely tooled to perform their function. The attainment of an early, benign immortality will be no less complex in terms of medical intervention. But there are now good reasons for thinking that it is fundamentally possible.

ACKNOWLEDGMENTS

I am grateful to L.D. Mueller, J.P. Phelan, and C.L. Rauser for their comments on an earlier draft.

References

1) Rose, MR; 1991: *Evolutionary Biology of Aging*; Oxford University Press; pg.17–

2) Martinez, DE; "Mortality patterns suggest a lack of senescence in Hydra" in: *Experimental Gerontology* (1998, Vol. 33); pg. 217–

3) Bell, G; "Evolutionary and nonevolutionary theories of senescence" in: *American Naturalist* (1984, Vol. 124); pg.600–

4) Greenwood, M & Irwin, JO; "Biostatistics of Senility" in: *Human Biology* (1939, Vol. 11); pg.1–

5) Gavrilov, LA, & Gavrilova, NS; *The Biology of Lifespan: A Quantitative Approach* (1991); Harwood

6) Carey, JR & Liedo, P & Orozco, D & Vaupel, JW; "Slowing of mortality rates at older ages in large medfly cohorts.", in: *Science* (1992, Vol. 258); pg.457–

7) Curtsinger, JW & Fukui, HH & Townsend, DR & Vaupel, JW; "Demography of genotypes: failure of the limited life-span paradigm in Drosophila melanogaster." in: *Science* (1992, Vol. 258); pg.461–

8) Fukui, HH & Xiu, L & Curtsinger, JW; "Slowing of age-specific mortality rates in Drosophila melanogaster", in: *Experimental Gerontology* (1993, Vol. 28); pg.585–

9) Vaupel, JW & Carey, JR & Christensen, K & Johnson, TE & Yashin, A I & Holm, NV & Iachine, IA & Kannisto, V & Khazaeli, AA & Liedo, P & Longo, VD & Zeng, Y & Manton, KG & Curtsinger, J W; "Biodemographic trajectories of longevity" in: *Science* (1998, Vol. 280); pg.855–

10) Drapeau, MD & Gass, EK & Simison, MD & Mueller, LD & Rose, MR; "Testing the heterogeneity theory of late-life mortality plateaus by using cohorts of Drosophila melanogaster." in: *Experimental Gerontology* (2000, Vol 35); pg.71–

11) Charlesworth, B; *Evolution in Age-Structured Populations*, (1980); Cambridge University Press

12) Mueller, LD & Rose, MR; "Evolutionary theory predicts late-life mortality plateaus." in: *Proceedings of the National Academy of Sciences USA* (1996, Vol. 93); pg.15249–

13) Charlesworth, B & Partridge, L; "Aging: Leveling of the grim reaper." in: *Current Biology* (1997, Vol. 7); pg.R440–

14) Pletcher, SD & Curtsinger, JW; "Mortality plateaus and the evolution of senescence: why are old-age mortality rates so low?" in: *Evolution* (1998, Vol. 52); pg.454–

15) Charlesworth, B; "Patterns of age-specific means and genetic variances of mortality rates predicted by the mutation-accumulation theory of ageing." in: *Journal of Theoretical Biology* (2001, Vol. 210); pg.47–

16) Shaw, B; *One Million Tomorrows* (1970); Ace Books

17) Rose, MR & Drapeau, MD & Yadzi, PG & Shah, KH & Moise, DB & Thakar, RR & Rauser, CL & Mueller, LD; "Evolution of late-life mortality in Drosophila melanogaster." in: *Evolution* (2002,Vol. 56); pg.1982–

The War on Aging

Speculations on Some Future Chapters in the Never-Ending Story of Human Life Extension

Aubrey de Grey, Ph.D.

Until the 1800's, over a quarter of those born, even in wealthy nations, died before their first birthday and a huge number of women died in childbirth. The great French doctor Louis Pasteur can therefore be credited without much argument as the person who has extended more lives by more years than anyone in history, by virtue of his introduction of the germ theory and the consequent appreciation of the importance of hygiene in medical care and, subsequently, of the power of antibiotics. [1] (Had the medical establishment been less robust in their suppression of new ideas, Pasteur would have been preceded by over a decade by Ignaz Semmelweis [2]; I return to this point towards the end of this essay.) This insight resulted, in the industrialized world, in a reduction by an order of magnitude in infant mortality over a period of a few decades. [3] It would doubtless have arrived eventually, but we have no idea how long it might have been delayed had Pasteur not existed. If we guess at a ten-year delay, which is less than what Semmelweis's suppression caused, we can conclude that tens of millions of people who would have died before the age of one instead lived to an age typical of those of their cohort who survived infancy – certainly at least 50 years on average. Thus, Pasteur may well have added on the order of a billion person-years to human life.

The demographic result is that deaths in infancy are now vastly outnumbered by deaths between ages 50 and 100. But this is not reflected in our priorities, as demonstrated by the resources allocated to medical research and care. Enormous effort is made to maintain the life of sick babies, and nobody disputes the merit of such a policy. Indeed, it seems difficult to imagine any argument against it that does not utterly fly in the face of all that we instinctively know about human morality.

In this article I explore an extremely straightforward scenario, whose neglect by others can, I feel, stem only from unjustifiable short-sightedness: that humanity's progress in reducing death rates at older ages will recapitulate the sequence just outlined for infant mortality. Some time – quite possibly within only a few decades, as I have discussed extensively elsewhere and will therefore only summarize here – we will make breakthroughs in maintaining and restoring the health and vigor of the elderly comparable, in terms of healthy years added to their lives, to what Pasteur and those who implemented his ideas gave to those otherwise destined to die in infancy. Forever thereafter, I suggest, we will strive vigorously to reduce the incidence of involuntary death (at whatever age) yet further. In the sections that follow I sketch some of the major advances that we seem likely to make in this endeavor. The later episodes that I describe may seem uninterestingly distant at first, but such nonchalance becomes questionable when one considers how early in this chain of events those with access to the latest medical care will begin to enjoy a diminishing mortality risk – an increasing *remaining* life expectancy as time passes. (I like to call this the achievement of 'life extension escape velocity'.) In a nutshell, I claim it is probable that most of the first generation of 150-year-olds (defined as those who reach 150 and are aged at most 30 years younger than the first 150-year-old) – a group who are almost certainly already alive and may well be middle-aged – will not die unless at their own hand.

The Second Worst Thing that Can Happen

Anti-aging medicine worthy of the name does not yet exist and seems certain not to exist for at least 15–20 years. By "worthy of the name" I mean interventions that can reliably restore someone exhibiting age-related dysfunction to the physiological and cognitive robustness that they enjoyed in early adulthood.

This fact is apparently such grim reading for those unlikely to survive long enough to see a cure for aging, that society allows the term "anti-aging medicine" to be used for products that have no discernible efficacy in even slowing aging down, let alone reversing it. [4] Such people have an alternative to the certainty of permanent oblivion, however: oblivion that may well be permanent but may, just possibly, be only temporary. Cryopreservation is catching on, slowly but surely: around 1000 individuals are signed up to have their heads (and sometimes also their bodies) immersed in liquid nitrogen upon their clinical and legal death. [5] This may sound like very few, but not when we consider how resistant the general public remains to the idea of serious life extension by purely biomedical means (even as a goal, let alone a foreseeable one). [6] In fact, I surmise that cryonics sign-ups are a respectable and probably increasing proportion of that small group who embrace the goal of extreme life extension – and this is hardly surprising, given the simplicity of arguments such as Merkle's characteristically razor-sharp quotation: "Cryonics is an experiment. So far the control group isn't doing very well." [7] If so, the main reason people are not beating a path to cryonics organizations is not that they are unconvinced of the *feasibility* of being resuscitated in a post-aging era, but that they doubt its *desirability*. How long will this remain so? We will come back to the question of cryonics later.

I like to define the War On Aging (WOA) as the period beginning with the announcement of mammalian (almost certainly mouse) life extension results impressive enough to force public opinion to accept the foreseeability of serious human life extension, and ending with the release of effective human life extension therapies, albeit initially at a price that only the relatively wealthy can afford.

DEFEATING AGING AS WE KNOW IT TODAY AND AS IT WILL THEREBY BECOME

The War On Aging is only the precursor of the (indefinite) period of rapid and sustained reduction in mortality rates. While the WOA is in progress, there will be a sharp rise in the number of people who adopt lifestyle changes to improve their chances of surviving long enough to benefit from life extension therapies. I suspect, however, that these changes will be modest compared to what happens when those therapies actually appear. They will quite probably still be ample to destabilize many aspects of modern society – avoidance of risky but vital jobs being an obvious example – but the nagging acknowledgement that we still do not know how soon the human therapies will actually materialize (and, therefore, whether they will be in time for us) will limit the adoption of the more extreme measures typical of wartime as we know it.

The post-WOA era will begin when appreciable numbers of middle-aged people – let us say, the richest 10% of those in the richest 10 nations – have access to medical care that extends their healthy lifespan by at least two decades. What will such care comprise?

As I have discussed extensively elsewhere, [8–10] there seem to be only seven broad categories of molecular and cellular difference between older and younger people that we need to fix to achieve two decades of human life extension ("seven deadly things"). These consist of a decline in the number of cells in certain tissues and an accumulation of unwanted cells of certain types, of mutations in our chromosomes, of mutations in our mitochondria, of random cross-links between long-lived extracellular proteins and of chemically inert but bulky 'junk' in our lysosomes and in extracellular spaces.

Further, I have delineated [8;10–13] approaches to either repairing or obviating (stopping from being pathogenic however much they accumulate) all these changes. All these approaches are already technically feasible. The underlying precursor technologies have already been developed and the work needed to complete them can be described in considerable detail. I have termed these projects 'Strategies for Engineered Negligible Senescence' (SENS) [8–10], since their goal is collectively to eliminate from humans the positive correlation between age and risk of death per unit time – biogerontologists' formal definition of senescence.

Unfortunately, most of the first-generation SENS therapies will be not only risky and laborious but also partial. A thorough survey of this issue exceeds the scope of this essay, so I will discuss just one illustrative example here: the breaking of extracellular protein-protein crosslinks.

Most such links are laid down by a process called glycoxidation, in which proteins react with sugars in the circulation to form adducts that can rearrange and undergo subsequent, oxidative reactions forming linkage to a neighboring protein. [14] Such crosslinks are eventually harmful to long-lived extracellular structures, especially the artery wall, because they make these structures less elastic and thus more prone to mechanical damage. The most promising first-generation therapy to eliminate such cross-links is a molecule known as ALT-711. [15] However, it breaks only one class of such link, known as dicarbonyl bonds. [16] Numerous other classes have been identified. [17] Worse, many are thermodynamically much more stable than dicarbonyl bonds, giving reason to doubt that non-toxic small molecules could ever cleave them. More elaborate approaches (already under consideration, but beyond the scope of this article) may thus be needed in order to limit the abundance of crosslinks indefinitely.

The 'Escape Velocity' Transition

The above considerations constitute an acknowledgement that aging will never be cured in the sense that a bacterial infection is cured, i.e. entirely eliminated from the body. Rather, it will be cured in the sense that malaria or AIDS can presently be cured: it can be controlled very well given access to appropriate medication whenever needed, but that medication can never be confidently dispensed with forever. It may not be clear to the reader, however, that what I have said above allows even that degree of 'cure' of aging.

The reason it does so can be summed up in one word: bootstrapping. The second-generation therapies will not be needed for at least two decades after the first-generation therapies arrive, because that is how long it will take before the "seven deadly things" return to the life-threatening level that they attained before those therapies arrived. Thus, so long as the second-generation therapies do indeed arrive within two decades, we will be broadly in the same position as regards mortality rates, as if they had arrived at the same time as the first-generation. The same logic clearly extends to subsequent generations of therapies, without limit. Two decades is an eternity in science, especially in well-funded science (which life extension will certainly be at this time). Second-generation therapies are thus virtually certain to arrive in time. Hence, as soon as we reach the point of extending life expectancy by even a couple of decades, we can be confident that most beneficiaries of such therapies will survive to benefit from subsequent ones. Those people's life expectancy will thus be indefinite, even though they are still aging. The analogy with a projectile propelled from the Earth at greater than escape velocity is not perfect, but I find it evocative.

IMPROVING OUR LEAD-TIME

For the reasons surveyed above, I consider it appropriate to regard the WOA as ending when first-generation SENS therapies become widely available. Thus far, however, I have only discussed specific, identifiable problems – which are, necessarily, unimaginatively similar to the targets of the first-generation therapies. What about things we have not thought of?

I predict that this point will motivate – starting as the WOA nears its end – a research project that will dwarf even the WOA itself. Our unarguably limited ability to predict what aging will throw at us next could, it would seem, only be addressed reliably by clairvoyance. Or could it? Could we metaphorically press the fast-forward button to discover what the future holds? We are exceedingly fortunate that such an option is indeed available.

Specifically, I predict that humanity will create and maintain a very large colony of non-human primates of several different species – probably totaling tens of thousands of animals – on which to test novel life extension therapies. Primates have three characteristics which, jointly, motivate this action: They are biologically extremely similar to us; They don't talk, so if the biomedical imperative is sufficient society feels entitled to do more or less anything to them; They all age at least twice as fast as us.

Because of this, a large colony of primates maintained under conditions very similar to those under which we maintain ourselves – the same range of diets, the same lack of exercise, and of course the same medical care, including all life-extension treatments in use at the time – will be virtually certain to display any health-threatening characteristic of aging that we ourselves exhibit, at an age at most half that at which it appears in us. These primates will be the experimental recipients of succeeding generations of rejuvenation therapies. Some such therapies will have unforeseen side-effects that will kill some of the colony, which is why we will need such a large colony so as to maintain a sufficient number of them of an age sufficiently exceeding half the age of any human yet alive to ensure that our primate experiments succeed before their results are needed. Splendidly, this becomes progressively easier as time passes: we may only just have 80-year-old primates before we have 160-year-old humans, but we will certainly have 100-year-old primates some years before we have 200-year-old humans, and the lead-time improves forever thereafter. This strategy will be our most powerful defense against the unforeseeable biomedical challenges that our attainment of unprecedented ages will create.

AVOIDING INVOLUNTARY DEATH FROM ALL CAUSES

When I entered biogerontology I saw nothing very wrong, nothing undignified, about death; what I hated was aging. I wanted to let people live the lives they choose; if someone wished to live fast and thereby knowingly risk dying young, I saw nothing wrong in a world in which that person probably would indeed die young. In recent years, however, I have come to believe otherwise.

The principal basis for my change of heart is the stark incompatibility of my previous position with the way people with a respectable remaining life expectancy and an appreciation of it actually behave. Those most inclined to engage in life-threatening activities are the young, who have not fully grasped their own mortality and the underprivileged, whose remaining life expectancy is always modest on account of the lesser availability of medical care (especially preventative care), the higher incidence of violent crime, and so on. The same is, I believe, true at a global scale. Perhaps it is sheer luck that we are approaching the 60th anniversary of the last time that any western European nations were at war with each other or internally, an interval not previously seen since Roman times. But I strongly suspect that this arises from a sea change in the readiness of both policy-makers and their electorate to sacrifice large numbers of their own lives in the interests of national pride. The elimination of the death penalty throughout Europe and the increasingly stringent restrictions on firearms ownership seen in the UK are examples of the same phenomenon as is the increasing public hostility to the habit of driving under the influence of alcohol. The same process is occurring throughout the industrialized world, albeit lagging somewhat behind Europe in several respects. It is for this sort of reason – simple extrapolation from the past century – that I predict that society will act to ensure that death from 'extrinsic' causes remains a lot rarer than death from causes that physiologically young individuals usually escape. This will entail a considerable acceleration in the rate at which we alter our lifestyles. (I predicted in 1999 that once we cure aging driving will be outlawed [18]; I still think that is likely, at least unless cars become much more automated and accidents thereby made very rare even after severe human error.) This is the final component of the logic underlying my prediction [19] that the average age at death of those born in wealthy nations in the year 2100 will exceed 5000 years, which is perhaps five times the value resulting from a permanent enjoyment of the mortality rate of young teenagers in such nations today.

RETHINKING CORPOREAL CONTINUITY

One contributor to involuntary death was omitted from my survey in the preceding section: frustration. Adjustments in society to diminish greatly the incidence of death from armed conflict, homicide and accidents are already accepted as welcome in principle; all that is in question is the extent to which such measures can be implemented without unacceptably infringing human rights or expending resources that could achieve similar life extension in other ways, and these arguments become weaker as the amount of life lost by an avoidable death increases. It will thus be entirely in keeping with contemporary social norms if the cure of aging, and the widespread appreciation that violent death deprives the individual of an indefinite number of years, causes society to embrace such changes, even ones that were vigorously resisted hitherto. But the same cannot necessarily be said for lifestyle changes that severely and permanently impact the quality of our newly indefinite lives. Unfortunately, some of the activities that make our lives fulfilling are associated with a definite risk. Sometimes only a small risk, but even that may, at that time, be greater than our risk of dying of old age; and therefore become something we take much more seriously than we do today. Further, some such risks – notably, exposure to new infectious diseases – seem likely to yield non-uniformly and unpredictably, if at all, to the increased expenditure that will surely be marshaled against them. [20] Thus, our continuing and ever-improving success in avoiding physical or mental frailty will be at the increasingly unsatisfactory price of eschewing fulfilling activities that just might kill us.

For this reason, I expect that in centuries to come we will work to develop what might be termed "non-invasive static uploading" or, more prosaically, backing-up of our cognitive state. At this point it is plausible, though obviously not known, that all the moderately persistent components of that state (that is, everything except short-term memory) are encapsulated in the network of synaptic connections between our neurons. The strength of those connections probably matters too, but maybe not to all that much precision. It is also plausible that, a century or three from now, an extremely high-resolution version of magnetic resonance imaging may exist that can scan a living person's brain and detect all such information. This amount of data could easily be stored electronically by then, as it exceeds by only 3–5 orders of magnitude the storage capacity of today's personal computers. Further, cells in culture will by then be amenable not only to differentiation along desired lineages but to stimulation to form synaptic connections with particular other cells to which their axons are juxtaposed. This means that, in principle, a copy of a living person's brain – all trillion cells of it – could be constructed from scratch, purely by *in vitro* micromanipulation of neurons into a synaptic network previously scanned from that brain. (Doing this in less than geological time would be possible because it could be highly parallelized. Since most axons are extremely short, small parts of the new brain could be constructed in separate vessels, leaving only a small minority of connections to be made to link these modules together.) If this were performed at close to zero degrees Centigrade, no electrical activity would occur during construction; such a brain (after installation into a much more straightforwardly reconstructed body) could then be 'awakened' simply by slow warming, much as people often recover from hypothermia-induced comas. In this way, a human could be created with the cognitive (and, if necessary, an improvement on the physical) state of someone deceased, even in the absence (which I suspect will be with us for a very long time indeed) of a detailed understanding of how that cognitive state results from that synaptic network. It is the lack of need for such understanding that distinguishes this procedure from the transfer of our cognitive state to radically different hardware on which it would actually 'run', which is the essence of the full-blown 'uploading' concept introduced by Moravec. [21]

If so, who would the constructed person be? They would surely claim to be the deceased person. It would be difficult to disagree, because we already have a precedent: corporeal continuity is not the basis for our emotional attachment to the person who fell asleep in our bed last night or the one who will wake up in our bed tomorrow. Rather, we identify with that person because we know that their state of *mind* – irrespective of how many atoms their body shares with ours – was/will be so deeply similar to our own that they were/will be unarguably us. This line of thought plainly raises phenomenal philosophical questions about identity (in cases where someone is constructed from a backup of someone who is not dead, for example), but I suspect that such esotericisms will not long restrain us when we see it as a way out of the by then intolerably risk-averse character of our existence. This is the final step in my argument for supposing that even the first generation of sesquicentenarians will mostly not die involuntarily; their acquisition (by the means described in previous sections) of a 5000-year life expectancy means that, even if the scanning and reconstruction technology posited in this section takes 500 years to develop, most of them will still be alive – in a youthful state – to take advantage of it.

I now return to the topic of resuscitating cryonics patients. Much work has gone into developing technology to lower a person to liquid nitrogen temperatures without forming ice crystals in their cells, because such crystals decimate cell membranes and thereby render implausible the resuscitation of the individual in the future, even presuming sophisticated technology to address the cause of their death. [22] I think this may not have been as important as most have supposed, and I base this view on a consideration of the circumstances in which a cryonics patient is and is not likely to be devitrified. It will not be enough to have cryopreserved and resuscitated a chimpanzee, for example, and failed to detect any difference in its personality, because assays of that personality will be inadequate to reveal changes of a subtlety that would still matter if they occurred in a human. The choice to resuscitate will simply not be made while even a small risk is perceived that a resuscitation will be only a qualified success, and if technology can be foreseen (even distantly so) which would substantially diminish that risk. Hence, I strongly suspect that those currently residing in cryonic containers in Scottsdale and Detroit will be resuscitated by the scanning and reconstruction approach just outlined, and not by thawing or devitrifying their original body. And it seems highly likely that such a scan could be performed just as successfully on a brain shot through with ice crystals as on one that had been perfectly vitrified.

In Conclusion

The first-generation SENS therapies that will give middle-aged humans an extra 20–50 years of healthy life may be developed well before mid-century, or they may not. But however long it takes, most of the first cohort who benefit from those therapies will probably live as long as they choose to, whereas those only five or ten years older will only have the opportunity of cryonics to live beyond 150. Every day that we can expedite the development of SENS will, therefore, probably confer on roughly 100,000 people the opportunity to extend their life span indefinitely – and this figure is largely independent of when SENS arrives. We can no longer pretend that we know so little about how to cure aging that the timing of this advance will be determined overwhelmingly by future serendipitous discoveries: we are in the home straight already. We are therefore perpetrating, right now, an offence that puts the medical establishment's suppression of Semmelweis in the shade.

References

1) Schwartz, Maxime; "The life and works of Louis Pasteur" in: *Journal of Applied Microbiology* (2001, Vol. 91); pg. 597–

2) Carter, Codell K & Carter, Barbara R; *Childbed Fever: A Scientific Biography of Ignaz Semmelweis* (1994); Greenwood Publishing Group

3) Armstrong, Gregory L & Conn, Laura A & Pinner, Robert W; "Trends in infectious disease mortality in the United States during the 20th century" in: *Journal of the American Medical Association* (1999, Vol. 281); pg. 61–

4) Olshansky, Jay S & Hayflick, Leonard & Carnes, Bruce A; "Position statement on human aging" in: *Journals of Gerontology A Biological Sciences Medical Sciences* (2002, Vol. 57A); pg. B292–

5) Paggetti, Maria; "The new ice age" in: *Esquire* (2003, Vol. 139); no pagination available. Reproduced at http://www.cryonics-europe.org/esquire.htm

6) Miller, Richard A; "Extending life: scientific prospects and political obstacles" in: *Milbank Quarterly* (2002, Vol. 80); pg. 155–

7) Alcor Life Extension Foundation; *Notable Quotes*, (2003) http://www.alcor.org/notablequotes.html ; Originally quoted in the *Alcor Indiana Newsletter* (June 1992)

8) de Grey, Aubrey DNJ & Ames, Bruce N & Andersen, Julie K & Bartke, Andrzej & Campisi, Judith & Heward, Christopher B & McCarter, Roger JM & Stock, Gregory; "Time to talk SENS: critiquing the immutability of human aging" in: *Annals of the New York Academy of Sciences* (2002, Vol. 959); pg. 452–

9) de Grey, Aubrey DNJ & Baynes, John W & Berd, David & Heward, Christopher B & Pawelec, Graham & Stock, Gregory; "Is human aging still mysterious enough to be left only to scientists?" in: *BioEssays* (2002, Vol. 24); pg. 667–

10) de Grey, Aubrey DNJ; "An engineer's approach to the development of real anti-aging medicine" in: *Science of Aging Knowledge Environment* (2003, Vol. 2003); http://sageke.sciencemag.org/cgi/content/full/sageke;2003/1/vp1

11) de Grey, Aubrey DNJ; "Mitochondrial gene therapy: an arena for the biomedical use of inteins" in: *Trends in Biotechnology* (2000, Vol. 18); pg. 394–

12) de Grey, Aubrey DNJ; "Bioremediation meets biomedicine: therapeutic translation of microbial catabolism to the lysosome" in: *Trends in Biotechnology* (2002, Vol. 20); pg. 452–

13) de Grey, Aubrey DNJ & Campbell, F. Charles & Dokal, Inderjeet & Fairbairn, Leslie J & Graham, Gerry J & Jahoda, Colin AB & Porter, Andrew CG; "Total deletion of in vivo telomere elongation capacity: an ambitious but possibly ultimate cure for all age-related human cancers" in: *Annals of the New York Academy of Sciences* (2004, Vol. 1019); pg. 147–".; in press

14) Monnier, Vincent M & Cerami, Anthony; "Nonenzymatic browning in vivo: possible process for aging of long-lived proteins" in: *Science* (1981, Vol. 211); pg. 491–

15) Asif, Mohammad, Egan, John, Vasan, Sara, Jyothirmayi, Garikiparthy N, Masurekar, Malthi R, Lopez, Santos, Williams, Chandra, Torres, Ramon L, Wagle, Dilip, Ulrich, Peter, Cerami, Anthony, Brines, Michael, Regan, Timothy J; "An advanced glycation endproduct cross-link breaker can reverse age-related increases in myocardial stiffness" in: *Proceedings of the National Academy of Sciences of the United States of America* (2000, Vol. 97); pg. 2809–

16) Vasan, Sara; Zhang, Xin, Zhang, Xini, Kapurniotu, Aphrodite, Bernhagen, Jurgen, Teichberg, Saul, Basgen, John, Wagle, Dilip, Shih, David, Terlecky, Ihor, Bucala, Richard, Cerami, Anthony, Egan, John, Ulrich, Peter; "An agent cleaving glucose-derived protein crosslinks in vitro and in vivo" in: *Nature* (1996, Vol. 382); pg. 275–

17) Baynes, John W; "The role of AGEs in aging: causation or correlation" in: *Experimental Gerontology* (2001, Vol. 36); pg. 1527–

18) de Grey, Aubrey DNJ; *The mitochondrial free radical theory of aging* (1999); Landes Bioscience

19) Richel, Theo; "Will human life expectancy quadruple in the next hundred years? Sixty gerontologists say public debate on life extension is necessary" in: *Journal of Anti-Aging Medicine* (2003, Vol. 6); pg. 307–

20) Lashley, Felissa R; "Factors contributing to the occurrence of emerging infectious diseases" in: *Biological Research for Nursing* (2003, Vol. 4); pg. 258–

21) Moravec, Hans; *Mind Children* (1988); Harvard University Press

22) Wowk, Brian & Leitl, Eugen & Rasch, Christopher M & Mesbah-Karimi, Nooshin & Harris, Steven B & Fahy, Gregory M; "Vitrification enhancement by synthetic ice blocking agents" in: *Cryobiology* (2000, Vol. 40); pg. 228–

The Dream of Elixir Vitae

João Pedro de Magalhães, Ph.D.

Human aging is a universal process of loss of viability and increase in vulnerability. Although so far the underlying mechanisms of aging remain largely a mystery, it is reasonable to expect that we will eventually understand the human aging process. Possibly during this century, we will know what changes occur in a human being from ages 30 to 70 to increase the chance of dying by roughly 32-fold. Yet even if researchers detail those changes, even if researchers identify the causal molecular and cellular mechanisms responsible for human aging, this will not necessarily lead to a cure for aging. HIV was identified as the cause of AIDS over 20 years ago and we still cannot cure AIDS. [1] So to delay aging, not to mention to stop or reverse the human aging process, will be a monumental task. It is true that we still do not know in detail what changes occur as human beings age, but it is equally true that the most important question in studying aging is not why we age but how can we fix it.

THERAPY AS INFORMATION

A disease, any type of disease, is a time-dependent change in the body that leads to discomfort, pain, or even death. Therapies aim to delay, stop, or reverse those changes from occurring either by large-scale interventions, such as surgery, or by transmitting the necessary information to the body. For example, a bacterial infection may be reversed by penicillin, which is an information vector that 'tells' the bacterial wall to 'open', thus killing the bacteria and reversing the disease state. Most pharmaceutical interventions are, in essence, information vectors transmitting instructions that are intended to delay, stop, or reverse the time-dependent changes related to a given pathology. Antibiotics, pain-killers, corticosteroids, anti-depressants, and many more products fit this description. Yet present therapies transmit relatively simple instructions: a pain-killer 'tells' neurons to stop transmitting pain signals and corticosteroids 'tell' the immune system to diminish its response. Curing aging will most likely require the transmission of much larger amounts of information.

Aging is a sexually transmitted terminal disease that can be defined as a number of time-dependent changes in the body that lead to discomfort, pain, and eventually death. In order to cure aging we will need to target multiple types of cells and address different types of molecular damage and malfunction. That is why organ transplants and surgery will not be the solution for aging, at least not a definitive cure. The future of medicine is not in large-scale interventions but in smaller, less invasive but more precise therapies. The solution to aging is not in addressing individual age-related pathologies but rather in tiny structures that are able to instruct our body to become young again.

Thanks to the enzyme telomerase, it is possible to prevent cells in culture from certain forms of aging. [2] It is equally possible to reverse the genetic program of adult cells back to youthfulness by cloning techniques. [3] There is no law of nature to prevent us from instructing the cells of an adult human being to avoid aging by, for example, changing the genetic program at a DNA or epigenetic level. Since, like any disease, aging results from disrupted or unbalanced molecules it is also theoretically possible to reverse age-related changes by precise molecular and cellular therapies. [4;5]

To slow, stop, and reverse human aging we will likely require three steps: (1) remove damaged or inactive molecules and cells; (2) restore function to several molecules and cells by repair or replacement; (3) modify the genetic program to prevent the aging process from repeating itself. These interventions are what we will most likely need to balance the body's chemical reactions and molecular structural changes that become disrupted as we age. Yet how can we transmit such massive amounts of information to our body?

INSTRUCTING THE HUMAN BODY

Most pharmaceutical interventions are composed of chemicals or biomolecules usually transmitting a single signal to the body: acetyl-salicylic acid, also known as aspirin, the anti-depressant fluxetine, hormones, etc. Novel findings in chemical genetics may allow the development of small molecules that target specific genes and pathways. [6] Yet the simple instructions these deliver to our cells are unlikely to be enough to cure aging. Assuming that aging is, to a large degree, programmed in our genes [7], curing aging will require technologies that are not yet available. To give an example, there are dozens of inherited diseases originating in single genes for which there is no cure simply because we lack the technologies to turn on and off human genes. Since curing aging will require us to transmit large amounts of information to the body, new technologies will be necessary. Herein, I will first give a brief overview of the most promising technologies to address this problem: gene therapy and single-gene interventions, cell therapy and stem cells, and nanotechnology. Afterwards, I will attempt to foresee how we can cure aging based on these technologies and what breakthroughs will be necessary.

GENE THERAPY

Gene therapy has been hailed as a major tool to deliver information, genes in this case, to the human body. [8] Although genes can be injected directly [9], most gene therapy methods involve the use of a vector for the specific purpose of inserting DNA into cells. Viruses are the most widely used vector and several experiments have already shown the power of this technology. In one exciting discovery, virus-induced expression of IGF-1, a growth factor, reversed age-related changes in the skeletal muscle of mice. Increases of almost 30% in strength were witnessed in treated old animals when compared to controls. [10] If aging may be reversed by the expression of key genes, then gene therapy holds great promise. Neuronal death has also been delayed by the introduction of a single gene using the herpes virus [11] and reversal of age-associated neural atrophy was achieved in monkeys by gene therapy. [12]

Gene therapy is promising but limited in scope due to the inherited 'bandwidth' constrains of the technique. Large-scale genetic engineering is already possible in embryos [13] and maybe our grandchildren will be born without aging. But present-day gene therapy does not provide a technology to cure aging in adults. The main reason is that viruses cannot 'transport' much genetic information. A typical virus carries up to a few hundred thousand base pairs, which is meaningless when compared to the three billion base pairs of the human genome. Maybe it is possible to use a combination of viruses but there are other problems. Viral vectors can stably integrate the desired gene into the target cell's genome but the gene's integration may occur at oncogenes (cancer-inducing genes), causing cancer. An immune response against viruses or transgenes may also occur and could be fatal as in the famous case of Jesse Gelsinger. [14] Virus-based gene therapy does not appear adequate to cure aging for not only is its safety dubious but the amount of genetic information viruses can carry is insufficient.

In addition to viruses, it has also been proposed that certain bacteria can act as vectors in gene therapy – the major advantage being that bacteria can transport larger amounts of information and still be able to change the genome. [15] As with viral-induced gene therapy, the immune response is a major problem. Some promising results have emerged from cancer treatments [16] but it is dubious bacterial-based vectors can become a solution to aging within a near future due to safety concerns.

If gene therapy can be used to express certain genes, RNA interference or RNAi can be used to inactivate them. Tiny double-stranded molecules of RNA can be designed to block a given target gene. [17] For example, it has been proposed that blocking the action of the gene responsible for Huntington's disease may prevent the onset of this disease. RNAi can be seen as another type of information vector used to transmit information to the body. Of course there are limitations, but if specific genes have to be turned off at specific times to cure aging, RNAi appears a viable solution. For instance, oncogenes appear to be activated during aging. For these, RNAi and 'classical' single-molecule-based pharmaceutical interventions [18] appear a viable solution.

CELL THERAPY

Gene therapy and RNAi are limited in the number of genes they can affect in cells. One way to overcome this limitation is by replacing the cells themselves, a process known as 'cell therapy'. Since there are few theoretical restrictions as to the number of genetic modifications cells can endure, cell therapy has a greater 'bandwidth'. For example, in an experiment aimed at treating the immunodeficiency disease SCID-X1, cells from the immune system were extracted from a patient, genetically engineered, and inserted back again with encouraging results. [19]

One growing area involves stem cells. A stem cell is a sort of 'unprogrammed' cell that has the potential to become any type of cell in the adult body. Aging has been linked to an age-related inability of stem cells to replenish mature cells and so therapeutic interventions that enhance stem cell functional capacity might ameliorate the age-associated atrophies of several organ systems. [20] More importantly, with nuclear transfer experiments such as *Dolly* [3], it is now possible to 'turn back the clock' and generate embryonic stem cells from an adult. [21;22] In theory, it is possible to genetically modify these cells according to needs, differentiate them into the necessary tissue or organ and then implant them to treat age-related diseases, a procedure called 'therapeutic cloning'. [23;24] Since these cells are genetically identical to the patient's there are few or no problems of immune rejection.

The ability of stem cells to regenerate virtually all types of tissues holds great promise. [25] In theory, it is possible to create practically all components of a human being in the lab and then replace the patient's organs and tissues one by one. Stem cells have been used with success against heart disease, [26] or to repair damage to the brain [27] and spinal cord. [28] Also, stem cells are incredibly versatile: transplantation of mesenchymal stem cells into the bone marrow has shown that they can travel through the body and become bone or muscle cells where needed. [29] These experiments demonstrate how a few cells can impact on whole organs by fostering regeneration, how a few tiny cells can transmit massive amounts of information to the human body.

Although much research is necessary and stem cells are still too expensive for widespread use, the basics for using these techniques are known and we can expect more practical applications to emerge in a near future. The ability stem cells have to sprout regeneration and repair tissues makes them an excellent candidate for anti-aging therapies.

NANOTECHNOLOGY

An adult human, once a tiny cell, is a self-assembling machine made of trillions of microscopic components. Roughly, a human being consists of $\sim 7 \times 10^{27}$ atoms and $\sim 10^5$ different molecular species, mostly proteins [30]. Genes and proteins are organic nanostructures working with molecular precision to form complex components such as human cells. The concept of nanotechnology, first proposed by Richard Feynman and later developed by the pioneering work of Eric Drexler, is our ability to manipulate matter and energy at smaller scales (one billionth of a specified unit is called a 'nano'). This capacity will increase until we reach and surpass our own biological nanostructures [4;31]. One key concept in nanotechnology is the molecular assembler, a machine capable of assembling other molecules given a set of instructions and the necessary resources. Ribosomes, the sites where proteins are built based on the instructions of the genes, are known molecular assemblers. A man-made molecular assembler capable of building molecule-scale machines to guide specific chemical reactions would allow the construction of devices with atomic precision capable of a myriad of functions.

In theory, nanostructures can be built to drive chemical reactions capable of reversing aging by reversing chemical reactions and damage that occur as we age. The goal would be to build the necessary nanostructures to reverse age-related changes with minimal perturbation. For example, damage to DNA increases with age. Even though it is debatable whether this is a result or a cause of aging, it appears likely that if we could build nanostructures to reverse these changes it could reverse at least some aspects of age-related disease. The body already features several of these nanostructures as part of the DNA repair machinery. Enhancing it with novel nanostructures could help turn the balance of DNA damage versus repair in our favor and thus reverse this form of damage. The applications of nanotechnology are manifold and it is not possible to describe them all, but one possible application would be to design bacteria, viruses, or even stem cells to perform large-scale gene therapy without being attacked by the immune system. For example, by taking the viral nanostructures for integrating foreign DNA into host cells and apply them to stem cells. [32]

Nanotechnology holds great expectations and promises. The greatest problem is that, so far, nanotechnology is almost exclusively theoretical without any clinical or medical trials. Even so, nanomachines aimed at correcting molecular defects for which there is no 'natural' tool – e.g., removal of lipofuscin, also called age-pigment – may be necessary. [33]

Changing the Soul of Man

The ultimate aim of research on aging is to create what medieval alchemists called 'Elixir Vitae', what science-fiction writer David Zindell called 'Godseed' [34] an entity capable of reversing the molecular and cellular changes that occur as we age and changing the genome of our cells to prevent aging from happening again. Initially, the Elixir will need to transmit a signal to drive regeneration, as happens in apparently non-aging animals such as lobsters [35] and turtles. [36] It may even be the case that tissue regeneration will eliminate damaged molecules and quiescent cells while at the same time restoring function. Otherwise the Elixir will have to incorporate ways to eliminate nonfunctional nanostructures and cells while at the same time restoring youthful vigor. Afterwards, the regenerated tissue will need to be prevented from aging again, probably by including the necessary instructions together with the instructions ordering regeneration. [37].

From a technological perspective, the Elixir will likely be a combination of the techniques presented previously: a mix of RNAi, gene therapy, and stem cells. The goal is to instruct the body's cells to regenerate while suppressing undesired genes. In addition, even if we do not know in detail how to reverse all age-related changes and pathologies, we may address specific pathologies through conventional therapies. For instance, to rejuvenate the immune system we will need to prevent the thymus from degenerating and so specific interventions will be necessary. Eventually, novel nanostructures may allow us to reverse specific age-related degenerative changes. [32] Yet we will not need mature nanotechnology for building the Elixir. It is impossible to say if man-made molecular assemblers will emerge in 10, 50, or 500 years from now, so we should not, and need not, depend upon nanotechnology to cure aging. As such, the core of the Elixir will likely be stem cells.

One specific case is the brain, the source of our consciousness. Again, the primary strategy should be to foster regeneration. It appears dangerous to use viruses and bacteria as vectors for gene therapy in the brain, so again stem cells hold the greatest promise. Non-invasive methods to express exogenous genes in the brain already exist and may serve to express specific critical genes. [38]

In addition, several species such as reptiles, lobsters and birds feature advanced regenerative capacities and appear not to age. Deriving information from these species to engineer how to rebuild the human genome to avoid aging is also within our reach. [39] In another example, work is being conducted to attempt to implement the advanced regenerative capacity of amphibians to mammals. [40] Synthetic biology and information systems will be the 'glue' that binds all these fields together and allow us to design, regulate, and apply the Elixir.

Conclusion

Elixir Vitae needs not be anything besides present technologies combined with some engineering feats. Importantly, the theoretical basis for these technologies already exists. What remains is the engineering problem of making them work according to our needs. Namely, we must (1) develop therapies based on stem cells for tissue regeneration; (2) implement synthetic biology to control stem cells; (3) test and develop the safety and accuracy of RNAi, gene therapy, and molecular therapies; (4) learn more about regeneration and the signals involved in each type of tissue; (5) apply whole genome engineering to aging. Lastly, we need to know, of course, where to act. That is, what causes aging in humans, what makes us gradually weaker and more vulnerable – but that is not the subject of this article.

It remains that Elixir Vitae is not just a utopia but also an achievable goal that we can build, hopefully within a reasonable time span.

ACKNOWLEDGEMENTS

Thanks to the FCT, Portugal, for their generous financing. J.P. de Magalhães is supported by NIH-NHGRI CEGS grant to George Church. Further thanks to Robert Bradbury, George Church, and Aubrey de Grey for our wonderful and creative discussions.

References

1) Gallo, RC & Montagnier, L; "The discovery of HIV as the cause of AIDS" in: *New England Journal of Medicine* (2003, Vol. 349); pg.2283–

2) Bodnar, AG & Ouellette, M & Frolkis, M & Holt, SE & Chiu, CP & Morin, GB & Harley, C B & Shay, JW & Lichtsteiner, S & Wright, WE; "Extension of life-span by introduction of telomerase into normal human cells" in: *Science* (1998, Vol. 279); pg.349–

3) Wilmut, I & Schnieke, AE & McWhir, J & Kind, AJ & Campbell, KH; "Viable offspring derived from fetal and adult mammalian cells" in: *Nature* (1997, Vol. 385); pg.810–

4) Drexler, EK; *Engines of Creation* (1986); Anchor Press

5) de Grey, AD; "An engineer's approach to the development of real anti-aging medicine" in: *Science of Aging Knowledge Environment* (2003, Vol. 2003); pg.VP1

6) Schreiber, SL; "The small-molecule approach to biology: Chemical genetics and diversity-oriented organic synthesis make possible the systematic exploration of biology" in: *Chemical & Engineering News* (2003, Vol. 81); pg.51–

7) de Magalhaes, JP; "Is mammalian aging genetically controlled?" in: *Biogerontology* (2003, Vol. 4); pg.119–

8) Lyon, J & Gorner, P; *Altered Fates: Gene Therapy and the Retooling of Human Life* (1995); Norton

9) Symes, JF & Losordo, DW & Vale, PR & Lathi, KG & Esakof, DD & Mayskiy, M & Isner, JM; "Gene therapy with vascular endothelial growth factor for inoperable coronary artery disease" in: *Annals of Thoracic Surgery* (1999, Vol. 68); pg.830–

10) Barton-Davis, ER & Shoturma, DI. & Musaro, A & Rosenthal, N & Sweeney, HL; "Viral mediated expression of insulin-like growth factor I blocks the aging-related loss of skeletal muscle function" in: *Proceedings of the National Academy of Sciences of the United States of America* (1998, Vol. 95); pg.15603–

11) Antonawich, FJ & Federoff, HJ & Davis, JN; "BCL-2 transduction, using a herpes simplex virus amplicon, protects hippocampal neurons from transient global ischemia" in: *Experimental Neurology* (1999, Vol. 156); pg.130–

12) Smith, DE & Roberts, J & Gage, FH & Tuszynski, M H; "Age-associated neuronal atrophy occurs in the primate brain and is reversible by growth factor gene therapy" in: *Proceedings of the National Academy of Sciences of the United States of America* (1999, Vol. 96); pg.10893–

13) Chan, AW & Chong, KY & Martinovich, C & Simerly, C & Schatten, G; "Transgenic monkeys produced by retroviral gene transfer into mature oocytes" in: *Science* (2001, Vol. 291); pg.309–

14) http://www.circare.org/pg.htm

15) Theys, J & Barbe, S & Landuyt, W & Nuyts, S & Van Mellaert, L & Wouters, B & Anne, J & Lambin, P; "Tumor-specific gene delivery using genetically engineered bacteria" in: *Current Gene Therapy* (2003, Vol. 3); pg.207–

16) Thomas, CE & Ehrhardt, A & Kay, MA; "Progress and problems with the use of viral vectors for gene therapy" in: *Nature Reviews Genetics* (2003, Vol. 4); pg.346– ref 20) "Journal of Cellular Physiology"

17) Tuschl, T; "Expanding small RNA interference" in: *Nature Biotechnology* (2002, Vol. 20); pg.446–

18) Haseltine, W; "Regenerative Medicine: Systematic Application of Biotechnology, Bioengineering, Nanotechnology and Information Sciences for the Improvement of Human Health", talk at the 10th Congress of The International Association of Biomedical Gerontology, 19–23 September 2003

19) Cavazzana-Calvo, M & Hacein-Bey, S & de Saint Basile, G & Gross, F & Yvon, E & Nusbaum, P & Selz, F & Hue, C & Certain, S & Casanova, JL; *et al.*, "Gene therapy of human severe combined immunodeficiency (SCID)-X1 disease" in: *Science* (2000, Vol. 288); pg.669–

20) Donehower, LA; "Does p53 affect organismal aging?" in: *Journal of Cellular Physiologyl* (2002, Vol. 192); pg.23–

21) Cibelli, JB & Lanza, RP & West, MD & Ezzell, C; "The first human cloned embryo" in: *Scientific American* (2002, Vol. 286); pg.44–

22) Hwang, WS & Ryu, YJ & Park JH & Park, ES & Lee, EG & Koo, JM & Jeon, HY & Lee, BC & Kang, SK & Kim, SJ & Ahn, C & Hwang, JH & Park, KY & Cibelli, JB & Moon, SY; "Evidence of a pluripotent human embryonic stem cell line derived from a cloned blastocyst" in: *Science* (2004, Vol. 303); pg.:1669–

23) Cibelli, JB & Stice, SL & Golueke, PJ & Kane, JJ & Jerry, J & Blackwell, C & Ponce de Leon, FA & Robl, JM; "Cloned transgenic calves produced from nonquiescent fetal fibroblasts" in: *Science* (1998, Vol. 280); pg.1256–

24) Lanza, RP & Cibelli, JB & West, MD; "Human therapeutic cloning" in: *Nature Medicine* (1999, Vol. 5); pg.975–

25) Krause, DS & Theise, ND & Collector, MI & Henegariu, O & Hwang, S & Gardner, R & Neutzel, S & Sharkis, SJ; "Multi-organ, multi-lineage engraftment by a single bone marrow-derived stem cell" in: *Cell* (2001, Vol. 105); pg.369–

26) Orlic, D & Kajstura, J & Chimenti, S & Jakoniuk, I & Anderson, SM & Li, B & Pickel, J & McKay, R & Nadal-Ginard, B & Bodine, DM & et al; "Bone marrow cells regenerate infarcted myocardium" in: *Nature* (2001, Vol. 410); pg.701–

27) Bjorklund, A & Lindvall, O; "Self-repair in the brain" in: *Nature* (2000, Vol. 405); pg.892–

28) Liu, S & Qu, Y & Stewart, TJ & Howard, MJ & Chakrabortty, S & Holekamp, TF & McDonald, JW; "Embryonic stem cells differentiate into oligodendrocytes and myelinate in culture and after spinal cord transplanta-tion" in: *Proceedings of the National Academy of Sciences of the United States of America.* (2000, Vol. 97); pg.6126–

29) Horwitz, EM & Prockop, DJ & Fitzpatrick, LA & Koo, WW & Gordon, PL & Neel, M & Sussman, M & Orchard, P & Marx, JC & Pyeritz, RE & *et al.*; "Transplantability and therapeutic effects of bone marrow-derived mesenchymal cells in children with osteogenesis imperfecta" in: *Nature Medicine* (1999, Vol. 5); pg.309–

30) http://www.foresight.org/Nanomedicine/Ch03_1.html#Tab0301

31) Feynman, R 1959, "There's Plenty of Room at the Bottom: An invitation to Enter a New Field of Physics", talk at the Annual Meeting of the American Physical Society, 29 December 1959

32) Freitas, R; *Nanomedicine; Biocompatibility* (2003, Vol. IIA*)*; Landes Bioscience

33) Freitas, R; *Nanomedicine; Basic Capabilities* (1999, Vol. I); Landes Bioscience

34) Zindell, D; *Neverness* (1990); Voyager

35) Klapper, W & Kuhne, K & Singh, KK & Heidorn, K & Parwaresch, R & Krupp, G; "Longevity of lobsters is linked to ubiquitous telomerase expression" in: *FEBS Letters* (1998, Vol. 439); pg.143–

36) Font, E & Desfilis, E & Perez-Canellas, MM & Garcia-Verdugo, JM; "Neurogenesis and neuronal regeneration in the adult reptilian brain" in: *Brain, Behavior and Evolution* (2001, Vol. 58); pg.276–

37) for an illustration see www.imminst.org/book1/

38) Shi, N & Pardridge, WM; "Noninvasive gene targeting to the brain" in: *Proceedings of the National Academy of Sciences of the United States of America* (2000, Vol. 97); pg.7567–

39) de Magalhaes, JP & Toussaint, O; "How bioinformatics can help reverse engineer human aging" in: *Ageing Research Reviews* (2004, Vol. 3); pg.125–

40) Brockes, J. P & Kumar, A; "Plasticity and reprogramming of differentiated cells in amphibian regeneration" in: *Nature Reviews Molecular Cell Biology* (2002, Vol. 3); pg.566–

Therapeutic Cloning

Michael D. West, Ph.D.

On a warm, still summer's night in August 1999, I stood in an Indiana hospital intensive care unit and turned my head to look at the clock. It was nearly 2am, the dark and deep hours before the morning light, when most human deaths occur. My dear mother's heart raced at 140 beats per minute, but that was about to end. She was dying, the woman who had given me life. I had long devised a plan that I hoped would one day help her, a plan some 20 years in the making. It was a plan to profoundly intervene in the biology of human aging. But I must say my best efforts seemed impotent at that moment, staring into the icy face of death.

At my request, a nurse pinched her fingernail bed one more time with a hemostat, squeezing her tender fingertip with the force of a pair of pliers. She winced, though imperceptibly. That was enough for the attending physician. She ordered the respirator that periodically forced air into my mother's lungs to be turned off along with the intravenous dopamine that was driving her heart. My eyes were fixed on the monitors. Mom's chest flattened. Her heart at first maintained its steady rhythm of 140 beats per minute and then slowly began its descent, drifting downward like a falling leaf in autumn – 140, 125, 110, 100…

My mind flashed back to a fall day in 1960 when I was seven years old. My mother and I walked along the sidewalk, on our way to the corner store. Suddenly, from above, a red leaf began a slow descent from the top of a tree in front of us. The leaf fell among some bushes alongside the sidewalk and I stopped to pick it up. "Mom, look, a cocoon." There among the fallen leaves was a gray cocoon, as big as your thumb, woven between the stems of a branch. I snapped it off and on we went to the store.

When we got home, my mother propped the cocoon on a ledge near a frosted kitchen window and I forgot about it over the long Michigan winter months. Then one spring day, a miracle happened. My mother and I had just stepped out of the car and my sister came in running, screaming, "Hurry, you gotta see!" Running into the kitchen I stopped at the door in amazement. A spectacular moth sat perched on the windowsill, more colorful, larger, and more wonderful than anything I knew existed – six inches from wing to wing, and painted in deep velvety colors of the rainbow. The miracle of this immortal cycle of metamorphosis – egg, caterpillar, moth and back to egg again – never left this young boy's mind.

THE CYCLE OF LIFE

For millennia our ancestors were observant enough to recognize the profundity of the cycle of life, and the fact that there is a sense in which life is immortal. While it is true that the individual plant ages and dies, out of the sun-drenched soil of spring a resurrection of plant life occurs every year. And while the individual zebra dies, as far back as anyone can remember there have always been zebras, and they always wear stripes. In other words, there is an immortal substratum of life, a continuum that connects the generations – a cycle of life, an immortal cycle. The individual passes away, but there is a continuity of individuals. The ancients attributed the force of this continual renewal of life to the realm of the gods.

The ancient Egyptians witnessed this immortal cycle of renewal on the banks of the river Nile. They came to revere its permanence. Like the sun that dies every evening in the western sky, only to be reborn the following morning, so the life of the individual is a transient phenomenon, but the immortal cycle of life itself is unchanging. In the mind of the ancient Egyptian mythologist, the phenomenon of immortal renewal was more than just a scientific observation; it was the cornerstone of the meaning of life itself. It was (so they reasoned) the work of a god, and they called that god Osiris.

Osiris, often depicted with his face painted green to symbolize this force of immortal renewal, was the foundation of ancient Egyptian religion. Osiris not only escaped death and corruption himself, but, inasmuch as any of his disciples could learn the mystery of the path into immortality, he too could hope for an immortal renewal of life transcending death.

Immortal Cells

I think the ancient Egyptian philosopher would have marveled to know that from the dry desert sand, future scientists would learn to make clear glass, and then to mold that glass into lenses, and then to stack those lenses together to make telescopes to magnify the night sky, and microscopes to magnify the world too small for the unaided eye. The microscope allowed early biologists to peer into the cellular substructure of life, and by the mid-1800's, it was confidently asserted that the mechanism of animal reproduction was via cells, not some amorphous "life force." All life comes from pre-existing life, and all cells come from pre-existing cells. In other words, science had uncovered the force of immortal renewal. It was an invisible thread that connected the generations, a lineage of microscopic primordial cells.

The German scientist August Weismann clearly understood the implications of this observation. The cell theory implied that life on our planet today likely originated many millions of years ago from single-celled animals that were immortal. By immortal Weismann did not mean to imply that they could not be killed. Indeed, the struggle of the fittest implied that their less-fit cousins did indeed die. By 'immortal' Weismann meant only that they need not die – that given proper nutrition, and barring some accident, any particular cell could continue dividing, leaving no dead ancestors in its wake.

Weismann then suggested that these original immortal cells may have clung to their daughter cells after dividing, thereby forming a small cluster of identical cells. It is then easy to imagine that these cells simply surrounded themselves with daughter cells to aid in their competition for immortality. One could imagine, for instance, that by "holding hands" in this manner, they were better able to move through the water, or perhaps better able to avoid being eaten by some other animal.

SPECIALIZATION OF CELLS

But complex, multicellular animals like you and me do leave dead ancestors behind. When and why did that happen? Here is where Weismann made a revolutionary proposal. He surmised that some of the cells in this cluster changed in a profound manner. When the largest animal was still a small cluster of cells – perhaps something like the ball of cells called Volvox, the microscopic pond water animal – some of these primordial and immortal cells specialized in a subtle way to facilitate the reproduction of their sister cells. These specialized cells, which are called 'somatic' cells (from the Greek word 'soma', meaning body), lost the ability to create other organisms like themselves. They had irreversibly specialized.

For the first time in history, a specialization of cell types arose. The change may have made the entire organism more fit compared to its competition, but the cost was that the somatic cells were destined to die, losing the potential for their own immortality. This, Weismann argued, was the first time programmed death appeared. As Joseph Wood Krutch (1856) put it:

> The amoeba and the paramecium are potentially immortal... But for Volvox, death seems to be as inevitable as it is in a mouse or in a man. Volvox must die, as Leeuwenhoek was to die because it had children and is no longer needed. When its time comes it drops quietly to the bottom and joins its ancestors. As Hegner, the Johns Hopkins zoologist, once wrote, "this is the first advent of inevitable natural death in the animal kingdom and all for the sake of sex. [1]

The question of the actual mechanisms of aging has been one of the most challenging questions mankind has ever faced. Weismann himself, recognizing the significance of this question, carefully considered the possible mechanisms of the body's aging. In 1881, he delivered a lecture to his fellow scientists at the Association of German Naturalists called 'Über die Dauer des Lebens', or 'On the Duration of Life'. It was the first effort to uncover the mechanisms of aging of the multicellular animal utilizing the sciences of cell biology and evolution. [2]

Let us now consider how it happened that the multicellular animals and plants, which arose from unicellular forms of life, came to lose this power of living forever. The answer to this question is closely bound up with the principle of division of labor the first multicellular organism was probably a cluster of similar cells, but these units soon lost their original homogeneity: the single cluster would come to be divided into two groups of cells, which may be called somatic and reproductive. As the complexity of the metazoan body increased, these two groups became more sharply separated from each other. Very soon the somatic cells surpassed the reproductive in number, and during this increase they became more and more broken up by the division of labor into sharply separated systems of tissues. As these changes took place, the power of reproducing large parts of the organism was lost, while the power of reproducing the whole individual became concentrated in the reproductive cells alone. But, it does not therefore follow that the somatic cells were compelled to lose the power of unlimited cell reproduction.

So, Weismann made the astonishing prediction that while the germ-line cells of multicellular animals, such as humans, were immortal (specifically, they could replicate without limit), the somatic cells were in fact mortal – that is, they had the capacity to divide only a finite number of divisions:

> Death takes place because a worn-out tissue cannot forever renew itself, and because a capacity for increase by means of cell-division is not everlasting, but finite. [2]

HAYFLICK'S EXPERIMENT

In 1961, the cell biologist Leonard Hayflick published the seminal work that convinced the scientific community that cells in the human body – the somatic cells – are mortal. [3] They could divide and proliferate, but as Weismann had predicted so many years earlier, even with optimum growth conditions they always eventually exhausted this ability and arrested their growth.

When I entered the field of aging research in the late 1970's, Hayflick's observation was already dogma. Humans are an amalgam of cells, some mortal and others immortal. Everyone is painfully aware of the mortal ones. Like bricks that are mortared side by side to construct the walls of buildings, so our cells are cemented together to form the tissues of our bodies. And those tissues – our bones, blood, and skin and the cells from which they are made – are all destined to age. We are made of mortal stuff. Our body's cells and therefore our bodies themselves share a common sentence of death. So, it may surprise you to learn that there is an exception.

HEIRS OF OUR IMMORTAL LEGACY

Still resident in the human body are potential heirs of our immortal legacy, cells that have the potential to leave no dead ancestors; cells from a lineage called the germ-line. These cells have the ability for immortal renewal as demonstrated by the fact that babies are born young, and those babies have the potential to someday make their own babies, and so on, forever.

In 1997, we at Geron Corporation, along with a host of collaborators, finally succeeded in isolating the gene that we reasoned should impart this capacity for unlimited replication in germ-line cells. The gene encodes a protein called telomerase that rewinds the clock of aging at the ends of the chromosome. The isolation of this 'immortality gene' stirred considerable controversy as to its potential to 'rewind' the Hayflick clock in cells in the human body after we showed that it actually works on cells cultured in a laboratory dish. Introducing the gene in an active state literally stops cellular aging. The cells become immortal but are still otherwise normal. This procedure, sometimes referred to as telomerase therapy, may indeed one day provide a means of transferring some of the powers of immortal renewal into at least some of the cells of the body. But it has proven difficult to efficiently introduce this, or indeed any gene, into most tissues in the human body.

STEM CELLS

And so, in the meantime, my mind turned to other ways to mine the rich vein of gold of the immortal germ-line. One fall day several years earlier, I took a break from working on telomerase and walked along the San Francisco Bay waterfront. I began thinking about what are called stem cells. A stem cell is a cell that can branch like the stems of a tree, either making another stem cell or changing to become a more specialized cell. There are all kinds of stem cells in the body, some more "potent" than others (that is, some have the potential to become more similar in cell type than others do).

I wondered that day whether it would be possible to grow a human totipotent stem cell in the laboratory. A human totipotent stem cell, though entirely theoretical at the time, could potentially branch into any cell in the body. If we imagine the branching of the fertilized egg cell into all the cells in the body, these totipotent stem cells would be analogous to the trunk of the tree of cellular life, the mother of all stem cells.

I was well aware of Weismann's work from my years working on cellular aging, and it occurred to me that if we could isolate and culture such cells from the human germ-line, they might be naturally immortal and telomerase positive, at least until they are directed to become a specific mortal cell type. And, most important of all, all the cells that come from them would be young, just as babies are born young.

EMBRYONIC STEM CELLS

In the following years and through the hard work of collaborators such as Jamie Thomson of the University of Wisconsin at Madison and John Gearhart at Johns Hopkins School of Medicine, the cells were finally isolated. They are called human embryonic stem cells because they come from human pre-implantation embryos (microscopic balls of cells that have not yet begun to develop and attach to the uterus to begin pregnancies.) These cells have fulfilled their promise in displaying the awesome power of making any cell type in the human body. And as we hoped, they made young cells that could theoretically be used to repair or replace aged or diseased cells and tissues.

President George W. Bush addressed the American people on August 9, 2001, to describe his policy relating to human embryonic stem cell research. He suggested that all federal funding be limited to the number of cell lines that had been isolated as of that date. He expressed his moral concerns about further efforts to isolate the cells, stating his religious belief that the entities from which the cells were derived were not in fact simply a clump of unformed cells, but instead were in fact very small people. [4]

There are several problems with the President's position. The practical one is that even if federal funding led to our ability to efficiently manufacture some cells of great therapeutic value, they would not be available to you – that is, the body would in most cases reject the transplanted cells as being a foreign invader. The miracle in the laboratory could not easily lead to a comparable miracle in the hospital bed.

THERAPEUTIC CLONING

And so in 1999, my colleagues and I proposed a controversial solution. We argued that the procedure called cell nuclear transfer – the transfer of a somatic cell into an enucleated egg cell – not only could produce embryos that when transferred into a uterus could produce a clone, but could also be harnessed to make embryonic stem cells as well. [5] Such cells would be essentially identical to the patient's cells. This could potentially solve the remaining problem of histocompatibility by creating human embryonic stem cells and then any cell in the body, all of which should never be rejected by the patient.

The use of somatic cell nuclear transfer for the purposes of reversing time's arrow on a patient's cells has been designated therapeutic cloning. This terminology is used to differentiate this clinical indication from the use of nuclear transfer for the cloning of a child, which in turn is often designated reproductive cloning.

Since the debate over therapeutic cloning began, the power of the technique has become increasingly impressive. In April 2000, my colleagues and I reported evidence that the egg cell could act as a "cellular time machine", not only reversing the arrow on differentiation (that is, not only converting a body cell like a skin cell into an embryonic stem cell), but also doing the unimaginable, returning the aged body cell to immortality and rewinding the clock of cellular aging as well. [6] These results, now reported for multiple mammalian species, suggest that we may have the potential to reverse the aging of human cells in the same manner.

This would mean that we could make young cells of any kind for a patient of any age. While this "time machine" is expected only to be big enough to take on a single cell, the resulting regenerated cells could theoretically be expanded and turned into cells that repopulate our blood with young immune cells, or cells that can re-seed our blood vessels with fresh young cells, or indeed young cells of any kind to treat a vast array of currently untreatable diseases.

HEATED CONTROVERSY

Despite the good intentions of researchers in this emerging field of regenerative medicine, these technologies have been at the center of one of the most heated controversies in the history of science. The raging controversy over embryonic stem cells and cloning has deeply divided our nation, and the stem cells' profound implications for battling the manifestations of age-related degenerative disease have raised concerns that mankind may be meddling in technologies that will anger the gods themselves.

In the face of lives molded and bounded by death, we are forced to choose our own position on these new technologies. In the summer of 1999, as I stood with my mother in a small hospital room, I knew my position in the debate. I would do anything to save the life of my mother – anything, that is, short of harming an actual human being.

And I had strong reasons to believe that therapeutic cloning would not have to create an individualized human being, even at the earliest states of development. I would risk my life, my finances, my reputation; I would give anything to help her.

DEATH IS THE ENEMY

My mother's pulse continued its downward glide – 90, 80, 20, 10, 8... I thought to myself, her heart was solid; I would never have worried that she would have died of heart failure.

I saw in my mind the swelling imbalance in blood chemistry, the millions of cells in her body screaming for help, her precious mind being turned to chaos by anoxia. Finally her heart cells – facing, for the first time since the origin of life on earth, the abyss of death – gave up their valiant defense of life and fell into chaos and arrhythmia. They had accomplished their appointed goal; they had successfully passed on their genome into a son. Minutes passed. As successful as my mother's life might have been in completing the job of reproduction, I found the strategy of the life cycle completely unacceptable. I stood there, hating death.

I walked to my car later that night, wandering aimlessly into the darkness. I had no itinerary, no plane reservations; I felt like driving randomly into the night. I looked overhead in that warm summer sky and stared at a bright but waning moon and I recognized its significance. The moon has for millennia been a source of encouragement to mankind facing the bleak realities of death and of loss. In 14 days, it is cut into pieces like the death of Osiris, but it always regenerates in an eternal fugue.

In the years to come, science and medicine will deliver on the promise of regenerative medicine. It is inevitable that the immortal cell, which can do so much to alleviate human suffering, will find its way to the hospital bed. But when these new therapies are available for our loved ones entirely depends on how we as a society grapple with these important issues.

The United States has a proud history of leading the world in boldly exploring new technologies. We did not hesitate to apply our best minds in an effort to enable a man to walk on the moon. We were not paralyzed by the fear that we would anger the gods by reaching for the heavens. But a far greater challenge stands before us now. We have been given two talents of gold. The first, the root of immortal human life, is the human embryonic stem cell. The second is nuclear transfer technology. Shall we, like the good steward of the Bible, take these gifts to mankind and courageously use them to the best of our abilities to alleviate the suffering of our fellow human beings, or will we fail most miserably and bury these gifts in the earth?

I am confident that the United States, which historically has led the way in advancing technology, will find the courage to lead in regenerative medicine as well. I only hope we will do so quickly; time is not on our side.

References

1) cited from Gilbert, Scott F.; *Developmental Biology*, (1991) Sinauer; pg. 13–18

2) Weismann, August; „Über die Dauer des Lebens" in: *The germ-plasm: a theory of heredity* (1882); Gustav Fischer Verlag; See Weismann, Augst, Translated by W. Newton Parker & Harriet Rionnfeldt; (1912) Charles Scribner's Sons

3) Hayflick, L & Moorhead, PS; "The serial cultivation of human diploid cell strains" in: *Experimental Cell Research* (1961, Vol. 253); pg. 585–621

4) http://www.whitehouse.gov/news/releases/2001/08/2001 0809-2.html

5) Lanza, RP & Cibelli JB & West MD; "Prospects for the use of nuclear transfer in human transplantation" in: *Nature Biotechnology* (1999, Vol. 17); pg. 1171–4

6) Lanza, RP & Cibelli, JB & Blackwell, C & Cristofalo, VJ & Francis, MK & Baerlocher, GM & Mak, J & Schertzer, M & Chavez, EA & Sawyer, N & Lansdorp, PM & West, MD; "Extension of cell life-span and telomere length in animals cloned from senescent somatic cells" in *Science;* (2000, Vol. 288); pg. 665–9

Nanomedicine

The quest for accident-limited healthspans

Robert A. Freitas Jr., J.D.

Editors Note: Dr. Freitas' original document includes a great number of graphics, tables and statistics that, for technical reasons, could not be included in this print version. You can obtain these graphics as well as an extended version of this article free of charge by visiting the E-reference site for this book:

<u>http://imminst.org/book1</u>

The greatest advances in halting biological aging and preventing natural death are likely to come from the fields of biotechnology and nanotechnology – that is, from nanomedicine. Nanomedicine is most simply and generally defined as the preservation and improvement of human health, using molecular tools and molecular knowledge of the human body [1].

Soon molecular tools of nanomedicine will include biologically active materials with well-defined nanoscale structures, such as dendrimer-based organic devices and pharmaceuticals based on fullerenes and organic nanotubes. We should also see genetic therapies and tissue engineering becoming more common in medical practice, which can contribute a little to life extension at the oldest ages.

In the mid-term, the next 5 or 10 years or so, knowledge gained from genomics and proteomics will make possible: (a) new treatments tailored to specific individuals, (b) new drugs targeting pathogens whose genomes have now been decoded, (c) stem cell treatments to repair damaged tissue, replace missing function, or slow aging, and (d) biological robots made from bacteria and other motile cells that have had their genomes re-engineered and re-programmed. We could also see artificial organic devices that incorporate biological motors or self-assembled DNA-based structures for a variety of useful medical purposes. We may even begin to see targeted anti-aging treatments which address each of the seven specific forms of cellular damage that produce pathologies leading to natural death, as described by Aubrey de Grey and colleagues [2], although there remain many institutional obstacles to direct progress via this conventional approach. [3]

In the farther term, perhaps somewhere in the 10 or 20-year time frame, the first fruits of molecular nanorobotics should begin to appear in the medical field. My own theoretical work in nanomedicine has concentrated on medical nanorobotics using diamondoid materials and nanoparts. This area, though clinically the most distant and still mostly theoretical, holds the greatest promise for health and life extension. With medical nanorobotics, we will gain the technological ability to perform specific internal repairs on individual cells in real time, thus largely eliminating all major causes of natural biological death.

The early theoretical work done by Drexler and Merkle, including most prominently a collection of bearings, gears, and other possible nanorobot parts, is well known. Possibly their most complex design was a nanoscale neon pump consisting of over 6,000 atoms, which was later simulated by computational chemists at California Institute of Technology. [5] The device could serve either as a pump for neon gas atoms or (if run backwards) as a motor to convert neon gas pressure into rotary power. The researchers reported that preliminary molecular dynamics simulations of the device showed that it could indeed function as a pump, although "structural deformations of the rotor can cause instabilities at low and high rotational frequencies". The motor was not particularly energy efficient – but it worked.

The ultimate goal of molecular nanotechnology is to develop a manufacturing technology able to inexpensively manufacture most arrangements of atoms that can be specified in molecular detail. Building medical nanorobots, each made of millions or billions of atoms, cheaply enough to be practical for medical therapies requires a new kind of manufacturing technology. Molecular manufacturing will be the ultimate mechanized production in terms of its precision and flexibility. Two central mechanisms have been proposed to achieve these goals at the molecular scale: (1) programmable positional assembly including, for example, fabrication of diamond structures, using molecular feedstock, and (2) massive parallelism of all fabrication and assembly processes.

As machine structures become more complex, getting all the parts to spontaneously self-assemble in the right sequence is increasingly difficult. To build complex structures, it makes more sense to design a mechanism that can assemble a molecular structure by what is called 'positional assembly' – that is, picking and placing molecular parts. A device capable of positional assembly at the molecular scale would work much like the robot arms that manufacture cars on automobile assembly lines in Detroit, or which insert electronic components onto computer circuit boards with blinding speed in Silicon Valley. Using the positional assembly approach, the robot manipulator picks up a part, moves it to the workpiece, and installs it. The robot then repeats the procedure over and over with many different parts until the final product is fully assembled.

In order to build durable nanorobots, we first must be able to fabricate parts made of diamond, sapphire, or similar strong materials. The controlled addition of carbon atoms to a growth surface of the diamond crystal lattice is called diamond mechanosynthesis. [6;7] In 2003, we proposed a new family of mechanosynthetic tools intended to be employed for the placement of two carbon atoms – a CC "dimer" – onto a growing diamond surface at a specific site. [6] These tools should be stable in vacuum and should be able to hold and position a CC dimer in a manner suitable for positionally controlled diamond mechanosynthesis at liquid nitrogen temperatures and possibly even at room temperatures. The function of a dimer placement tool is to position the dimer, then bond the dimer to a precisely chosen location on a growing diamond molecular structure, and finally to withdraw the tool, leaving the dimer behind on the growing structure. The diamond structure is then built up, dimer by dimer, until a complete molecularly precise nanopart has been fabricated.

Both the fabrication of nanoparts and the assembly of nanoparts into working nanorobots must be automated and must employ massive parallelism to be practical. There must be many hands at work simultaneously. Without this parallelism, there would be too many atoms per device (millions/billions) and too many devices needing to be assembled per application (trillions). New techniques for massively parallel positional assembly are being developed, including massively parallel manipulator arrays and self-replicating systems. One example of parallel assembly arrays, called "exponential assembly", has been proposed and patented by Zyvex [8]. There have also been many proposals for self-replicating systems known as "molecular assemblers" – tiny machines that could manufacture nanorobots with molecular precision [9].

What sorts of medical nanorobots could we build, and what would they do, if we could build them? The first simple device that I designed 9 years ago was the respirocyte, an artificial red blood cell [10].

Natural red cells carry oxygen and carbon dioxide throughout the human body. We have about 30 trillion of these cells in all our blood. Half our blood volume is red cells. Each red cell is about 3 microns thick and 8 microns in diameter. Respirocytes are much smaller than red cells, only 1 micron in diameter, about the size of a bacterium. Respirocytes are self-contained nanorobots built of 18 billion precisely arranged structural atoms. Each device has an onboard computer and an onboard powerplant.

I show them blue in color, because part of the outermost shell is made of sapphire, a tough ceramic made of aluminum and oxygen atoms that is almost as hard as diamond. These tanks could be safely charged up to 100,000 atmospheres of pressure, but we are conservative and only run them up to 1000 atmospheres. Most importantly, molecular pumps are arranged on the surface to load and unload gases from the pressurized tanks. Tens of thousands of individual pumps, called molecular sorting rotors, cover a large fraction of the hull surface of the respirocyte. As the rotor turns, molecules of oxygen (O_2) or carbon dioxide (CO_2) may drift into their respective binding sites on the rotor surface and be carried into (or out of) the respirocyte interior. There are 12 identical pumping stations laid out around the equator of the respirocyte, with oxygen rotors on the left, carbon dioxide rotors on the right, and water rotors in the middle. Temperature and concentration sensors tell the devices when to release or pickup gases. Each station has special pressure sensors to receive ultrasonic acoustic messages, so doctors can tell the devices to turn on or off, or change their operating parameters, while the nanorobots are inside a patient. The shaded area at left is the O_2

storage tank, the area at right is the CO_2 tank, the black dot at the center is the computer, and the open volume around the computer can be a vacuum, or can be filled or emptied with water. This allows the device to control its buoyancy very precisely and provides a crude but simple method for removing respirocytes from the blood using a centrifuge.

When we can build respirocytes, they could be used as an emergency treatment at the scene of a fire, where the victim has been overcome by carbon monoxide poisoning. In an animation [11] from the PBS documentary *Beyond Human*, 5 cubic centimeters of respirocyte-containing fluid are injected into the patient's vein. After passing through the pulmonary bed, the heart, and some major arteries, the respirocytes make their way into smaller, and smaller, blood vessels. After about 30 seconds, they reach the patient's capillaries and begin releasing life-giving oxygen to starving tissues. In the tissues, oxygen is pumped out of the device by the sorting rotors on one side. Carbon dioxide is pumped into the device by the sorting rotors on the other side, one molecule at a time. Half a minute later, when the respirocyte reaches the patient's lungs, these same rotors reverse their direction of rotation, recharging the device with fresh oxygen and dumping the stored CO_2, which can then be exhaled by the patient.

Only 5ccs of respirocytes, just 1/1000th of our total blood volume, could duplicate the oxygen-carrying capability of the entire human blood mass. Each respirocyte transports hundreds of times more physiologically available oxygen molecules than an equal volume of natural red blood cells. A half a liter of respirocytes, the most that could possibly be safely added to our blood, would allow a person to hold his breath at the bottom of a swimming pool for up to 4 hours, or to sprint at top Olympic speed for up to 12 minutes, without taking a breath.

Another medical nanorobot I designed more recently is the microbivore – an artificial white cell [12–15].

One main task of natural white cells is to absorb and digest microbial invaders in the bloodstream. This is called phagocytosis. Microbivore nanorobots would also perform phagocytosis, but would operate much faster, more reliably, and under human control. Like the respirocyte, the microbivore is much smaller than a red blood cell but is more complex than the respirocyte, having about 30 times more atoms involved in its construction.

The microbivore device is a flattened sphere with the ends cut off. It measures over 3 microns in diameter along its major axis and 2 microns in diameter along its minor axis. This size helps to ensure that the nanorobot can safely pass through even the narrowest of human capillaries and other tight spots in the spleen (e.g., the interendothelial splenofenestral slits [16]) and elsewhere in the human body. The microbivore has a mouth with an irising door, called the ingestion port, into which microbes are fed to be digested. The microbivore also has a rear end, or exhaust port. This is where the completely digested remains of the pathogen are expelled from the device. The rear door opens between the main body of the microbivore and a tail-cone structure. Inside the microbivore, there are two concentric cylinders. The bacterium is minced into little pieces in the morcellation chamber, the smaller inner cylinder, and then the remains are pushed into the digestion chamber, the larger outer cylinder. In a preprogrammed sequence engineered digestive enzymes are added, then removed, using an array of sorting rotors. In just 30 seconds these enzymes reduce the microbe's remains to simple chemicals like amino acids, free fatty acids, and simple sugars, which are then expelled harmlessly from the device. A human neutrophil, the most common type of leukocyte or white cell, can also capture and engulf a microbe in a minute or less, but complete digestion and excretion of the bug's remains can take an hour or longer.

But the first thing a microbivore has to do is reliably acquire a pathogen to be digested. If the correct bacterium bumps into the nanorobot surface, reversible binding sites on the microbivore hull can recognize and weakly bind to the bacterium. A set of 9 different antigenic markers should be specific enough, since all 9 must register a positive binding event to confirm that a targeted microbe has been caught. There are 20,000 copies of these 9-marker receptor sets, distributed in 275 disk-shaped regions across the microbivore surface. Inside the receptor ring are more rotors to absorb glucose and oxygen from the bloodstream for nanorobot power. At the center of each receptor disk is a grapple silo; each disk is 150 nanometers in diameter.

Once a bacterium has been captured by the reversible receptors, telescoping grapples rise up out of the microbivore surface and attach to the trapped bacterium. The microbivore grapples are modeled after a watertight manipulator arm originally designed by Drexler [17] for nanoscale manufacturing. This arm is about 100 nanometers long and has various rotating and telescoping joints that allow it to change its position, angle, and length. But the microbivore grapples need a greater reach and range of motion, so they are longer and more complex, with many additional joints. After rising out of its silo, a grapple arm can execute complex twisting motions, and adjacent grapple arms can physically reach each other, allowing them to hand off bound objects as small as a virus particle. Grapple handoff motions can transport a large rod-shaped bacterium from its original capture site forward into the mouth of the microbivore device. The bug is rotated into the proper orientation as it approaches the open mouth.

Our natural white cells – even when aided by antibiotics – can sometimes take weeks or months to completely clear bacteria from the bloodstream. By comparison, a single terabot dose of microbivores should be able to fully eliminate bloodborne pathogens in just minutes or hours, even in the case of locally dense infections. Microbivores would be up to ~1000 times faster acting than natural leukocytes. They'd digest almost 100 times more microbial material than an equal volume of natural white cells could digest, in any given time period.

Even more powerful applications – most importantly, involving cellular replacement or repair – are possible with medical nanorobotics. For example, most diseases involve a molecular malfunction at the cellular level, and cell function is significantly controlled by gene expression of proteins. As a result, many disease processes are driven either by defective chromosomes or by defective gene expression. So in many cases it may be most efficient to extract the existing chromosomes from a diseased cell and insert fresh new ones in their place. This procedure is called "chromosome replacement therapy".

During this procedure, your replacement chromosomes are first manufactured to order, outside of your body, in a clinical benchtop production device that includes a molecular assembly line. Your individual genome is used as the blueprint. If the patient wants, acquired or inherited defective genes could be replaced with nondefective base-pair sequences during the chromosome manufacturing process, thus permanently eliminating any genetic disease – including conditions related to aging. Nanorobots called chromallocytes [18], each carrying a single copy of the revised chromosomes, are injected into the body and travel to the target tissue cells. Following powered cytopenetration and intracellular transit to the nucleus, the chromallocytes remove the existing chromosomes and then install the properly methylated replacement chromosomes in every tissue cell of your body (requiring a total dose of several trillion nanorobots), then exit the cell and tissue, re-enter the bloodstream, and finally eliminate themselves from the body through the kidneys.

DECHRONIFICATION: A TREATMENT FOR THE DISEASE OF NATURAL DEATH

The end result of all these nanomedical advances will be to enable a process I call 'dechronification' – or, more colloquially, 'rolling back the clock'. I see no serious ethical problems with this. According to the volitional normative model of disease that is most appropriate for nanomedicine [19], if you're physiologically old and don't want to be, then for you, oldness and aging –and natural death – are a disease and you deserve to be cured. Dechronification will first arrest biological aging, then reduce your biological age by performing three kinds of procedures on each one of the 4 trillion tissue cells in your body:

First, a respirocyte – or microbivore – device will be sent to enter every tissue, to remove accumulating metabolic toxins and undegradable material. Afterwards, these toxins will continue to slowly re-accumulate as they have all your life, so you'll probably need a whole-body cleanout to prevent further aging, maybe once a year.

Second, chromosome replacement therapy can be used to correct accumulated genetic damage and mutations in every one of your cells. This might also be repeated annually, or less often.

Third, persistent cellular structural damage that the cell cannot repair by itself such as enlarged or disabled mitochondria can be reversed as required, on a cell-by-cell basis, using cellular repair devices. We're still a long way from having complete theoretical designs for many of these machines, but they all appear possible in theory. By the time our molecular manufacturing capability progresses to the degree necessary to begin building medical nanorobots, probably in the next 10–20 years, we will have good designs for cell repair devices.

The net effect of these interventions will be the continuing arrest of all biological aging, along with the reduction of current biological age to whatever new biological age is deemed desirable by the patient, severing forever the link between calendar time and biological health. These interventions may become commonplace, several decades from today.

Using annual checkups and cleanouts, and some occasional major repairs, your biological age could be restored once a year to the more or less constant physiological age that you select. I see little reason not to go for optimal youth – though trying to maintain your body at the ideal physiological age of 10 years old might be difficult and undesirable for other reasons. A rollback to the robust physiology of your late teens or early twenties would be easier to maintain and much more fun. That would push your Expected Age at Death up to around 700–900 calendar years. You might still eventually die of accidental causes, but you'll live ten times longer than you do now.

How far can we go with this? Well, if we can eliminate 99% of all medically preventable conditions that lead to natural death [19], your healthy life span, or health span, should increase to about 1100 years. It may be that you'll find it hard to coax more than a millennium or two out of your original biological body, because deaths from suicides and accidents have remained stubbornly high for the last 100 years, falling by only one-third during that time. But our final victory over the scourge of natural biological death, which we shall achieve later in this century, should extend the health span of normal human beings by at least ten- or twenty-fold beyond its current maximum length.

One can hope that the rate of suicides might be greatly reduced, with so much to look forward to, and with new nanomedical treatments for debilitating mental states becoming available. Nanotechnology can also improve the overall safety of our material environment – e.g., by making possible virtually crash-free, crash-safe cars and aircraft, buildings (including houses) that incorporate active safety devices, advanced nanomedicine for severe trauma anticipation and recovery, and the like – leading to vastly fewer deaths from accidents. Finally, genetic modifications or nanomedical augmentations to the human body [20] may extend healthy life spans still further, to a degree that cannot yet be accurately predicted.

References

1) Freitas RA Jr; "Section 1.2.2 Volitional Normative Model of Disease," in: *Nanomedicine, Volume I: Basic Capabilities, Landes Bioscience* (1999); pg. 18–20 http://www.nanomedicine.com/NMI/1.2.2.htm

2) de Grey, AB & Ames, BN & Andersen, JK & Bartke, A & Campisi, J &. Heward, CB & McCarter, RJ & Stock, G; "Time to talk SENS: critiquing the immutability of human aging," in: *Annals of the New York Academy of Sciences* 959 (2002); pg. 452–462, 463–465 // de Grey, AD & Baynes, JW & Berd, D & Heward, CB & Pawelec, G & Stock, G; "Is human aging still mysterious enough to be left only to scientists?" in: *Bioessays* 24 (2002); pg. 667–676, *Bioessays* 25 (2003); pg. 93–95 (discussion) // de Grey, AD; "An engineer's approach to the development of real anti-aging medicine," in: *Sci. Aging Knowledge Environment.* 2003 (2003):VP1 // de Grey, AD; "Challenging but essen-

tial targets for genuine anti-aging drugs," *Expert Opinion Therapeutic Targets* 7 (2003); pg. 1–5

3) Miller, Richard A; "Extending life: scientific prospects and political obstacles," in: *Milbank Quarterly* 80 (2002); pg.155–74 // de Grey, AD; "The foreseeability of real anti-aging medicine: focusing the debate," *Experimental Gerontology* 38 (2003); pg. 927–934

4) Freitas, Jr. Robert A; "Section 2.4.1 Molecular Mechanical Components," in: *Nanomedicine, Volume I: Basic Capabilities, Landes Bioscience* (1999), pg. 61–64 http://www.nanomedicine.com/NMI/2.4.1.htm

5) Cagin, T & Jaramillo-Botero, A & Gao, G & Goddard III, WA; "Molecular mechanics and molecular dynamics analysis of Drexler-Merkle gears and neon pump," in: *Nanotechnology* 9 (1998); pg. 143–152

6) Merkle, Ralph C & Freitas Jr., Robert A; "Theoretical analysis of a carbon-carbon dimer placement tool for diamond mechanosynthesis," in: *Journal of Nanoscience and Nanotechnology* 3 (2003); pg. 319–324 http://www.rfreitas.com/Nano/JNNDimerTool.pdf

7) Peng, Jinping & Freitas Jr., Robert A & Merkle, Ralph C; "Theoretical Analysis of Diamond Mechanosynthesis. Part I. Stability of C2 Mediated Growth of Nanocrystalline Diamond C(110) Surface," in: *Journal of Computational and Theoretical Nanoscience* 1 (2004) // Mann, David J & Peng, Jingping & Freitas Jr., Robert A & Merkle, Ralph C; "Theoretical Analysis of Diamond Mechanosynthesis. Part II. C2 Mediated Growth of Diamond C (110) Surface via Si/Ge-Triadamantane Dimer Placement Tools," in: *Journal of Computational and Theoretical Nanoscience* 1 (2004)

8) Merkle, Ralph C & Parker, Eric G & Skidmore, George D; "Method and system for self-replicating manufacturing stations," in: *United States Patent* No. 6,510,359, 21 January 2003

9) Freitas Jr., Robert A & Merkle, Ralph C; *Kinematic Self-Replicating Machines* (2004) Landes Bioscience; In press

10) Figure 15. An artificial red cell – the respirocyte [41]. Designer Robert A. Freitas Jr. ©1999 Forrest Bishop. Used with permission. http://www.foresight.org/Nanomedicine/Gallery/Species/Respirocytes.html

11) Lawrence Fields & Jillian Rose; "Animation of a respirocyte (an artificial red blood cell) being injected into the bloodstream," in: *PBS documentary "Beyond Human,"* air date 15 May 2001, Phlesch Bubble Productions website; http://www.phleschbubble.com/album/beyondhuman/respirocyte01.htm

12) A brief summary description may be found at: Robert A. Freitas Jr., "Microbivores: Artificial Mechanical Phagocytes," Foresight Update, No. 44, 31 March 2001, pg. 11–13; http://www.imm.org/Reports/Rep025.html.

13) The full technical paper is at: Freitas Jr., Robert A; "Microbivores: Artificial Mechanical Phagocytes using Digest and Discharge Protocol," in: *Zyvex preprint* (2001)

14) Freitas Jr. Robert A; website; http://www.rfreitas.com/Nano/Microbivores.htm

15) Images available at Nanomedicine Art Gallery, Foresight Institute website: http://www.foresight.org/Nanomedicine/Gallery/Species/Microbivores.html

16) Freitas Jr., Robert A; "Section 15.4.2.3 Geometrical Trapping in Spleen Vasculature," in: *Nanomedicine*, Volume IIA: Biocompatibility, Landes Bioscience, (2003); pg. 95–97 http://www.nanomedicine.com/NMIIA/15.4.2.3.htm#p6

17) Drexler, Eric K; "Section 13.4.1 A bounded-continuum design for a stiff manipulator," in: *Nanosystems: Molecular Machinery, Manufacturing, and Computation* (1992) edited by John Wiley & Sons; pg. 398–407

18) Freitas Jr., Robert A; "Chromallocytes: Cell Repair Nanorobots for Chromosome Replacement Therapy," (2004); in preparation.

19) Robert A. Freitas Jr., "Section 1.2.2. Volitional Normative Model of Disease," in: *Nanomedicine, Volume I: Basic Capabilities, Landes Bioscience* (1999), pg. 18–20; http://www.nanomedicine.com/NMI/1.2.2.htm

20) Freitas Jr., Robert A & Phoenix, Christopher J & "Vasculoid: A personal nanomedical appliance to replace human blood," in: *Journal of Evolution and Technology 11* (2002); http://www.jetpress.org/volume11/vasculoid.html

HUMAN BODY VERSION 2.0

Raymond Kurzweil, Ph.D.

In the coming decades, a radical upgrading of our body's physical and mental systems, already underway, will use nanobots to augment and ultimately replace our organs. We already know how to prevent most degenerative disease through nutrition and supplementation; this will be a bridge to the emerging biotechnology revolution, which in turn will be a bridge to the nanotechnology revolution. By 2030, reverse engineering of the human brain will have been completed and nonbiological intelligence will merge with our biological brains.

IT'S ALL ABOUT NANOBOTS

In a famous scene from the movie, *The Graduate*, Benjamin's mentor gives him career advice in a single word: "plastics". Today, that word might be "software", or "biotechnology", but in another couple of decades, the word is likely to be "nanobots". Nanobots – blood-cell-sized robots – will provide the means to radically redesign our digestive systems, and, incidentally, just about everything else.

In an intermediate phase, nanobots in the digestive tract and bloodstream will intelligently extract the precise nutrients we require, call for needed additional nutrients and supplements through our personal wireless local area network, and send the rest of the food we eat on its way to be passed through for elimination.

If this seems futuristic, keep in mind that intelligent machines are already making their way into our blood stream. There are dozens of projects underway to create blood-stream-based 'biological microelectromechanical systems' (bioMEMS) with a wide range of diagnostic and therapeutic applications. BioMEMS devices are being designed to intelligently scout out pathogens and deliver medications in very precise ways.

For example, a researcher at the University of Illinois at Chicago has created a tiny capsule with pores measuring only seven nanometers. The pores let insulin out in a controlled manner but prevent antibodies from invading the pancreatic Islet cells inside the capsule. [1] These nanoengineered devices have cured rats with type I diabetes, and there is no reason that the same methodology would fail to work in humans. Similar systems could precisely deliver dopamine to the brain for Parkinson's patients, provide blood-clotting factors for patients with hemophilia, and deliver cancer drugs directly to tumor sites. A new design provides up to 20 substance-containing reservoirs that can release their cargo at programmed times and locations in the body.

Kensall Wise, a professor of electrical engineering at the University of Michigan, has developed a tiny neural probe that can provide precise monitoring of the electrical activity of patients with neural diseases. Future designs are expected to also deliver drugs to precise locations in the brain. Kazushi Ishiyama at Tohoku University in Japan has developed micromachines that use microscopic-sized spinning screws to deliver drugs to small cancer tumors. [2]

A particularly innovative micromachine developed by Sandia National Labs has actual microteeth with a jaw that opens and closes to trap individual cells and then implant them with substances such as DNA, proteins or drugs. [3] There are already at least four major scientific conferences on bioMEMS and other approaches to developing micro- and nano-scale machines to go into the body and bloodstream.

Ultimately, the individualized nutrients needed for each person will be fully understood (including all the hundreds of phytochemicals from plants) and easily and inexpensively available, so we won't need to bother with extracting nutrients from food at all. Just as we routinely engage in sex today for its relational and sensual gratification, we will gain the opportunity to disconnect the eating of food from the function of delivering nutrients into the bloodstream.

This technology should be reasonably mature by the 2020's. Nutrients will be introduced directly into the bloodstream by special metabolic nanobots. Sensors in our blood and body, using wireless communication, will provide dynamic information on the nutrients needed at each point in time.

A key question in designing this technology will be the means by which these nanobots make their way in and out of the body. As I mentioned above, the technologies we have today, such as intravenous catheters, leave much to be desired. A significant benefit of nanobot technology is that unlike mere drugs and nutritional supplements, nanobots have a measure of intelligence. They can keep track of their own inventories, and intelligently slip in and out of our bodies in clever ways. One scenario is that we would wear a special 'nutrient garment' such as a belt or undershirt. This garment would be loaded with nutrient bearing nanobots, which would make their way in and out of our bodies through the skin or other body cavities.

DIGITAL DIGESTION

At this stage of technological development, we will be able to eat whatever we want, whatever gives us pleasure and gastronomic fulfillment, and thereby unreservedly explore the culinary arts for their tastes, textures, and aromas. At the same time, we will provide an optimal flow of nutrients to our bloodstream, using a completely separate process. One possibility would be that all the food we eat would pass through a digestive tract that is now disconnected from any possible absorption into the bloodstream.

This would place a burden on our colon and bowel functions, so a more refined approach will dispense with the function of elimination. We will be able to accomplish this using special elimination nanobots that act like tiny garbage compactors. As the nutrient nanobots make their way from the nutrient garment into our bodies, the elimination nanobots will go the other way. Periodically, we would replace the nutrition garment for a fresh one. One might comment that we do obtain some pleasure from the elimination function, but I suspect that most people would be happy to do without it.

Ultimately we won't need to bother with special garments or explicit nutritional resources. Just as computation will eventually be ubiquitous and available everywhere, so too will basic metabolic nanobot resources be embedded everywhere in our environment. In addition, an important aspect of this system will be maintaining ample reserves of all needed resources inside the body. Our version 1.0 bodies do this to only a very limited extent, for example, storing a few minutes of oxygen in our blood, and a few days of caloric energy in glycogen and other reserves. Version 2.0 will provide substantially greater reserves, enabling us to be separated from metabolic resources for greatly extended periods of time.

Once perfected, we will no longer need version 1.0 of our digestive system at all. I pointed out above that our adoption of these technologies will be cautious and incremental, so we will not dispense with the old-fashioned digestive process when these technologies are first introduced. Most of us will wait for digestive system version 2.1 or even 2.2 before being willing to dispense with version 1.0. After all, people didn't throw away their typewriters when the first generation of word processors was introduced. People held onto their vinyl record collections for many years after CDs came out (I still have mine). People are still holding onto their film cameras, although the tide is rapidly turning in favor of digital cameras. However, these new technologies do ultimately dominate, and few people today still own a typewriter. The same phenomenon will happen with our reengineered bodies. Once we've worked out the inevitable complications that will arise with a radically reengineered gastrointestinal system, we will begin to rely on it more and more.

Programmable Blood

As we reverse-engineer (learn the principles of operation of) our various bodily systems, we will be in a position to engineer new systems that provide dramatic improvements. One pervasive system that has already been the subject of a comprehensive conceptual redesign is our blood.

One of the leading proponents of 'nanomedicine', (redesigning our biological systems through engineering on a molecular scale) and author of a book with the same name is Robert Freitas, Research Scientist at the nanotechnology firm Zyvex Corp. Freitas' ambitious manuscript is a comprehensive road map to rearchitecting our biological heritage. One of Freitas' designs is to replace (or augment) our red blood cells with artificial 'respirocytes' that would enable us to hold our breath for four hours or do a top-speed sprint for 15 minutes without taking a breath. Like most of our biological systems, our red blood cells perform their oxygenating function very inefficiently, and Freitas has redesigned them for optimal performance. He has worked out many of the physical and chemical requirements in impressive detail.

It will be interesting to see how this development is dealt with in athletic contests. Presumably, the use of respirocytes and similar systems will be prohibited from Olympic contests, but then we will have the specter of teenagers in junior high school gymnasiums routinely outperforming Olympic athletes.

Freitas envisions micron-size artificial platelets that could achieve hemostasis (bleeding control) up to 1,000 times faster than biological platelets. Freitas describes nanorobotic micro-bivores (white blood cell replacements) that will download software to destroy specific infections hundreds of times faster than antibiotics, and that will be effective against all bacterial, viral and fungal infections, with no limitations of drug resistance.

Have a Heart, or Not

The next organ on my hit list is the heart. It's a remarkable machine, but it has a number of severe problems. It is subject to a myriad of failure modes, and represents a fundamental weakness in our potential longevity. The heart usually breaks down long before the rest of the body, and often very prematurely.

Although artificial hearts are beginning to work, a more effective approach will be to get rid of the heart altogether. Among Freitas' designs are nanorobotic blood cell replacements that provide their own mobility. If the blood system moves with its own movement, the engineering issues of the extreme pressures required for centralized pumping can be eliminated. As we perfect the means of transferring nanobots to and from the blood supply, we can also continuously replace the nanobots comprising our blood supply.

With the respirocytes providing greatly extended access to oxygenation, we will be in a position to eliminate the lungs by using nanobots to provide oxygen and remove carbon dioxide. One might point out that we take pleasure in breathing (even more so than elimination!). As with all of these redesigns, we will certainly go through intermediate stages where these technologies augment our natural systems, so we can have the best of both worlds. Eventually, however, there will be no reason to continue with the complications of actual breathing and the requirement of having breathable air everywhere we go. If we really find breathing that pleasurable, we will develop virtual ways of having this sensual experience.

We also won't need the various organs that produce chemicals, hormones, and enzymes that flow into the blood and other metabolic pathways. We already create bio-identical versions of many of these substances, and we will have the means to routinely create all biochemically relevant substances within a couple of decades. These substances (to the extent that we still need them) will be delivered via nanobots, controlled by intelligent biofeedback systems to maintain and balance required levels, just as our 'natural' systems do today (for example, the control of insulin levels by the pancreatic Islet cells). Since we are eliminating most of our biological organs, many of these substances may no longer be needed, and will be replaced by other resources that are required by the nanorobotic systems.

It is important to emphasize that this redesign process will not be accomplished in a single design cycle. Each organ and each idea will have its own progression, intermediate designs, and stages of implementation. Nonetheless, we are clearly headed towards a fundamental and radical redesign of the extremely inefficient and limited functionality of human body version 1.0.

So What's Left?

The skeleton is a stable structure, and we already have a reasonable understanding of how it works. We replace parts of it today, although our current technology for doing this has severe limitations. Inter-linking nanobots will provide the ability to augment and ultimately replace the skeleton. Replacing portions of the skeleton today requires painful surgery, but replacing it through nanobots from within can be a gradual and noninvasive process. The human skeleton version 2.0 will be very strong, stable, and self-repairing.

We will not notice the absence of many of our organs, such as the liver and pancreas, as we do not directly experience their functionality. The skin, however, is an organ we will actually want to keep, or at least we will want to maintain its functionality. The skin, which includes our primary and secondary sex organs, provides a vital function of communication and pleasure. Nonetheless, we will ultimately be able to improve on the skin with new nanoengineered supple materials that will provide greater protection from physical and thermal environmental effects while enhancing our capacity for intimate communication and pleasure. The same observation holds for the mouth and upper esophagus, which comprise the remaining aspects of the digestive system that we use to experience the act of eating.

REDESIGNING THE HUMAN BRAIN

The process of reverse engineering and redesign will also encompass the most important system in our bodies: the brain. The brain is at least as complex as all the other organs put together, with approximately half of our genetic code devoted to its design. It is a misconception to regard the brain as a single organ. It is actually an intricate collection of information processing organs, interconnected in an elaborate hierarchy, as is the accident of our evolutionary history.

The process of understanding the principles of operation of the human brain is already well under way. The underlying technologies of brain scanning and neuron modeling are scaling up exponentially, as is our overall knowledge of human brain function. We already have detailed mathematical models of a couple dozen of the several hundred regions that comprise the human brain.

The age of neural implants is also well under way. We have brain implants based on 'neuromorphic' modeling (i.e., reverse engineering of the human brain and nervous system) for a rapidly growing list of brain regions. A friend of mine who became deaf as an adult can now engage in telephone conversations again because of his cochlear implant, a device that interfaces directly with the auditory nervous system. He plans to replace it with a new model with a thousand levels of frequency discrimination, which will enable him to hear music once again. He laments that he has had the same melodies playing in his head for the past 15 years and is looking forward to hearing some new tunes. A future generation of cochlear implants now on the drawing board will provide levels of frequency discrimination that go significantly beyond that of 'normal' hearing.

Researchers at MIT and Harvard are developing neural implants to replace damaged retinas. [4] There are brain implants for Parkinson's patients that communicate directly with the ventral posterior nucleus and subthalmic nucleus regions of the brain to reverse the most devastating symptoms of this disease. An implant for people with cerebral palsy and multiple sclerosis communicates with the ventral lateral thalamus and has been effective in controlling tremors. "Rather than treat the brain like soup, adding chemicals that enhance or suppress certain neurotransmitters," says Rick Trosch, an American physician helping to pioneer these therapies, "we're now treating it like circuitry."

A variety of techniques are being developed to provide the communications bridge between the wet analog world of biological information processing and digital electronics. Researchers at Germany's Max Planck Institute have developed noninvasive devices that can communicate with neurons in both directions. [5] They demonstrated their 'neuron transistor' by controlling the movements of a living leech from a personal computer. Similar technology has been used to reconnect leech neurons and to coax them to perform simple logical and arithmetic problems. Scientists are now experimenting with a new design called 'quantum dots', which uses tiny crystals of semiconductor material to connect electronic devices with neurons. [6]

These developments provide the promise of reconnecting broken neural pathways for people with nerve damage and spinal cord injuries. It has long been thought that recreating these pathways would only be feasible for recently injured patients because nerves gradually deteriorate when unused. A recent discovery, however, shows the feasibility of a neuroprosthetic system for patients with long-standing spinal cord injuries. Researchers at the University of Utah asked a group of long-term quadriplegic patients to move their limbs in a variety of ways and then observed the response of their brains, using magnetic resonance imaging (MRI). Although the neural pathways to their limbs had been inactive for many years, the pattern of their brain activity when attempting to move their limbs was very close to that observed in non-disabled persons.

We will, therefore, be able to place sensors in the brain of a paralyzed person (e.g., Christopher Reeve) that will be programmed to recognize the brain patterns associated with intended movements and then stimulate the appropriate sequence of muscle movements. For those patients whose muscles no longer function, there are already designs for 'nanoelectromechanical' systems (NEMS) that can expand and contract to replace damaged muscles and that can be activated by either real or artificial nerves.

WE ARE BECOMING CYBORGS

We are rapidly growing more intimate with our technology. Computers started out as large remote machines in air-conditioned rooms tended by white-coated technicians. Subsequently they moved onto our desks, then under our arms, and now in our pockets. Soon, we'll routinely put them inside our bodies and brains. Ultimately we will become more nonbiological than biological.

The compelling benefits in overcoming profound diseases and disabilities will keep these technologies on a rapid course, but medical applications represent only the early adoption phase. As the technologies become established, there will be no barriers to using them for the expansion of human potential. In my view, expanding our potential is precisely the primary distinction of our species.

Moreover, all of the underlying technologies are accelerating. The power of computation has grown at a double exponential rate for all of the past century, and will continue to do so well into this century through the power of three-dimensional computing. Communication bandwidths and the pace of brain reverse-engineering are also quickening. Meanwhile, according to my models, the size of technology is shrinking at a rate of 5.6 per linear dimension per decade, which will make nanotechnology ubiquitous during the 2020's.

By the end of this decade, computing will disappear as a separate technology that we need to carry with us. We'll routinely have high-resolution images encompassing the entire visual field written directly to our retinas from our eyeglasses and contact lenses (the Department of Defense is already using technology along these lines from Microvision, a company based in Bothell, Washington). We'll have very-high-speed wireless connection to the Internet at all times. The electronics for all of this will be embedded in our clothing. Circa 2010, these very personal computers will enable us to meet with each other in full-immersion, visual-auditory, virtual-reality environments as well as augment our vision with location- and time-specific information at all times.

By 2030, electronics will utilize molecule-sized circuits, the reverse-engineering of the human brain will have been completed, and bioMEMS will have evolved into bioNEMS (biological nanoelectromechanical systems). It will be routine practice to have billions of nanobots (nano-scale robots) coursing through the capillaries of our brains, communicating with each other (over a wireless local area network), as well as with our biological neurons and with the Internet. One application will be to provide full-immersion virtual reality that encompasses all of our senses. When we want to enter a virtual-reality environment, the nanobots will replace the signals from our real senses with the signals that our brain would receive if we were actually in the virtual environment.

We will have panoply of virtual environments to choose from, including earthly worlds that we are familiar with, as well as those with no earthly counterpart. We will be able to go to these virtual places and have any kind of interaction with other real (as well as simulated) people, ranging from business negotiations to sensual encounters. In virtual reality, we won't be restricted to a single personality, since we will be able to change our appearance and become other people.

The most important application of circa-2030 nanobots will be to literally expand our minds. We're limited today to a mere hundred trillion inter-neuronal connections; we will be able to augment these by adding virtual connections via nanobot communication. This will provide us with the opportunity to vastly expand our pattern recognition abilities, memories, and overall thinking capacity as well as directly interface with powerful forms of nonbiological intelligence.

It's important to note that once nonbiological intelligence gets a foothold in our brains (a threshold we've already passed), it will grow exponentially, as is the accelerating nature of information-based technologies. A one-inch cube of nanotube circuitry (which is already working at smaller scales in laboratories) will be at least a million times more powerful than the human brain. By 2040, the nonbiological portion of our intelligence will be far more powerful than the biological portion. It will, however, still be part of the human-machine civilization, having been derived from human intelligence, i.e., created by humans (or machines created by humans) and based at least in part on the reverse-engineering of the human nervous system.

Stephen Hawking recently commented in the German magazine Focus that computer intelligence will surpass that of humans within a few decades. He advocated that we "develop as quickly as possible technologies that make possible a direct connection between brain and computer, so that artificial brains contribute to human intelligence rather than opposing it." Hawking can take comfort that the development program he is recommending is well under way.

References

1) Tao, Sarah & Dasai Tejal A; "Microfabricated Drug Delivery Systems: From particles to pores" in: *Advanced Drug Delivery Reviews* (2003, Vol. 55); pg.315–328

2) Jamieson, B & Buzsaki, G & Wise, KD; "A 96-Channel Silicon Neural Recording Probe with Integrated Buffers," in: *Annals of Biomedical Engineering*, (2000, Vol. 28 Supplement 1); pg. S-112

3) http://www.sandia.gov/media/NewsRel/NR2001/gobbler.htm

4) http://www.bostonretinalimplant.org/

5) Fromherz, Peter; "Neuroelectronic Interfacing: Semiconductor Chips with Ion Channels, Nerve Cells, and Brain" in: *Nanoelectronics and Information Technology* (2003) edited by Waser, R; Wiley-VCH Press; pg. 781–810

6) Winter, JO & Liu, TY & Korgel, BA & Schmidt, CE; "Recognition molecule directed interfacing between semiconductor quantum dots and nerve cells" in: *Advanced Materials* (2001, Vol. 13); pg. 1673–1677

Progress toward Cyberimmortality

William Sims Bainbridge, Ph.D.

Advances in information technology are essential for most of the imaginable means for achieving immortality, and fundamental to many. Before nanoscale robots are sent into a person's body to repair the damage from aging, computers will have to analyze what is needed and design the nanobots. [1;2] In the slow process of transferring a mind from an old brain into a freshly cloned one, that mind will need to be cached temporarily in an information system. This, then, raises the question of why it is necessary to transfer the mind from the information system into a vulnerable brain, rather than into a more durable robot or keeping it in the information system. [3]

Methods of Mind Reading

At a first approximation, there are two fundamental ways of reading the contents of a human mind into a computer: structural and functional. Each of these has innumerable variants that share a common principle.

In the structural approach, some process or device reads out the relevant structure of the brain and duplicates it inside a computer. The dominant structural assumption at present holds that a person's memories, mental skills, and much of the personality are encoded in the shape of the network of connections linking neurons. "Self lives in the synapses," might be the motto of this viewpoint. Actually, much smaller structures may also play a role, either inside the neurons or on their surfaces near the synapses. It has even been suggested that the glial cells, which outnumber neurons in the brain, are not merely supportive tissue but have some active function in thought or memory. Granting that further research is needed, let us assume for present purposes that a mind really does consist of the momentary structure of neuronal connections. At present, we can only imagine how that fine structure might be mapped.

Magnetic Resonance Imaging (MRI), Computer Assisted Tomography (CAT scans), Positron Emission Tomography (PET scans), Electroencephalography (EEGs or brain waves), and infrared observation are non-destructive ways of studying the brain. However, all of them have rather poor resolution. For example, MRIs are generally unable to resolve features smaller than a cubic millimeter, whereas thousands of neurons may jostle each other within this space. To see smaller structures requires increasing the power of the MRI scan, but this is dangerous and violates government regulations for research on human subjects. Perhaps computer analysis techniques can improve the resolution somewhat, but several of these methods already use sophisticated software, so we cannot count on really dramatic advances without a fresh approach. [4]

Unfortunately, at the present time it is difficult to see how the brain's detailed structure could be mapped without destroying it. In the *Visible Human Project* of the National Library of Medicine of the National Institutes of Health, two deceased human beings were frozen and then sliced so that their cross sections could be photographed, from head to toe. The images were then computerized so that three-dimensional models could be made of any of the organs. The resolution was at best a third of a millimeter, still far too gross to record the fine structure of the neurons, but the methods could probably be improved greatly. Whether through a similar mechanical sectioning process or through an intensive application of brain scan techniques, destructive mapping of the brain could conceivably chart the connections between the neurons. Storage and analysis of the data are well beyond current information technology capabilities.

Functional approaches have a different set of advantages and disadvantages, but they are ready today to make at least low fidelity copies of human personalities. While we can imagine many possibilities for the distant future, they will be of no value for the millions of people who will die before they can be developed.

If I had to make a prediction, I would guess that everyone alive when this essay is published will have died before structural methods of mind reading are perfected, and they will be of little value until a short time before then. This last point is based on the view that we will have to know very much about neural connections before we can deduce what meaning they represent. With functional mind reading, the meaning is attached to the data, so that even small fragments can be understood. A good metaphor is assembling a jigsaw puzzle when some parts of the image become perfectly clear long before we figure out what the whole picture is about.

Progress In Personality Capture

Functional mind reading is already possible at a moderate level of fidelity, and concentrated research efforts could achieve significant progress rapidly. Both traditional psychological tests and more recent computerized methods can collect a great deal of data rigorously about a person's skills, beliefs, behaviors, preferences, and emotional reactions.

My own research has focused on recording people's attitudes and preferences, building on decades of past work in such fields as sociology and political science that have become progressively computerized. [5–7] Attitudes are not merely personal but social, and my methodology begins with the ambient culture surrounding with the individual. [8] In May 1997, I launched a website called *The Question Factory* to create questionnaire items by posting open-ended surveys that asked respondents to write their views on various general topics. [9;10] For example, after pretesting on *The Question Factory* an open-ended question about what will happen over the coming century, I was able to place it in Survey2000, a massive web-based questionnaire sponsored by the National Geographic Society. About 20,000 people responded. From the several megabytes of predictions, I was able to edit 2,000 statements about the future that became fixed-choice questionnaire items, expressing the full range of views found in our culture. The respondent is supposed to say how likely it is that each idea will come true, and how good it would be if it did, so the resultant number of questions was actually 2 times 2,000 or 4,000. [11;12]

Other work with *The Question Factory* led to a total of 20,000 statements or 40,000 items. One set of 2,000 were stimuli that might elicit one of 20 different emotions in people: Anger, Boredom, Desire, Disgust, Excitement, Fear, Frustration, Gratitude, Hate, Indifference, Joy, Love, Lust, Pain, Pleasure, Pride, Sadness, Satisfaction, Shame, and Surprise. I then wrote a program for a pocket computer that would make it easy and convenient for a person to respond to a few items wherever they happened to be during the day. Each stimulus was rated in terms of how much it might produce each of the 20 emotions in a person, and in terms of three semantic differentials: Bad-Good, Weak-Strong, and Passive-Active, for a total of 46,000 responses. Naturally, I experiment on myself in this research, as well as with the help of other people, and to this point I have recorded my own personal answers to about 100,000 questions.

While I find answering questions fun, a kind of hobby, in general people will need to be motivated to record themselves, so future attitude recording systems will be designed to accomplish other things as well, such as psychotherapy and advice-giving. Recently, I have written a program called *ANNE* that runs on a small tablet computer that is easy to carry around. *ANNE* stands for ANalogies in Natural Emotions, is based on the 2,000 emotion stimuli, and helps the user orient emotionally to incidents that occur in life.

Suppose I have to give a lecture to a skeptical audience. I enter "give a lecture to a skeptical audience" into *ANNE*, then click a number of buttons (or speak my responses, because tablet computers have pretty good voice recognition systems), rating this incident on all the emotions and other variables. Immediately, *ANNE* compares my ratings with all the ratings of the 2,000 episodes stored in the computer, identifying which are most similar, so I can then contemplate what fundamental features they share and what strategies have worked well in the past in similar situations. Used over a period of years, *ANNE* would accumulate not only thousands of fresh episodes, but also much more information about my reactions to them than I would ordinarily remember – raising the paradox that a computer-generated duplicate of myself might be more like me than I am actually like myself at any given moment in my forgetful life.

Some quite prominent researchers in computer science or cognitive science are developing other means to record personality. From one perspective, we are what we experience. At Carnegie-Mellon University, from 1997 to 2000, Howard

Wactlar created a system called *Experience on Demand,* for the Defense Advanced Research Projects Agency, unobtrusively capturing people's experiences in a form that facilitates sharing them. [13] At Microsoft, Gordon Bell's team has been developing ways to collect and organize the documents and experiences of a lifetime in a project called *MyLifeBits.* [14;15]

Many researchers are developing computer methods to record how people perceive their environments. [16] Others are developing the technology not only to record real environments but to make virtual copies of them, notably the effort at Columbia University to duplicate electronically the Cathedral of Amiens, the Virtual Vaudeville project at the University of Georgia to recreate century-old performances like those of acrobat *Sandow the Magnificent,* and the *Monuments and Dust* project at the University of Virginia to recreate Victorian London, beginning with the famous Crystal Palace.

Recordings of behavior include facial expressions, [17] personal conversations [18] and the subtle delays when a person responds to challenging stimuli. [19;20] Lisa and Daniel Barrett [21] have used pocket computers to conduct a random sample of the things a person does or experiences, and we can well imagine that within a few years many people will have their wearable computers constantly switched on and sending their words, deeds, and feelings over wireless Internet to be recorded on a home digital library.

Artificial Intelligence

Once data about an individual personality have been ported into an information system, some method is needed to revive it. One common idea is that some form of artificial intelligence (AI) will reanimate the person's mind, so it is worth examining the current state of AI. In the past, proponents of AI oversold the capabilities of their technology, leading to a general stigmatization of the entire field within computer science. [22] Recently, however, there has been a resurgence of AI, coupled with a refocus.

People who availed themselves of the opportunity to visit the website of the National Science Foundation (NSF), could have seen an expansion of grant programs in this area over the past year or two. A new program in Computer Vision emerged from the program in Robotics and Human Augmentation. A new program in Human Language and Communication (really multi-modal computerized language processing) similarly emerged from the Human-Computer Interaction (HCI) program. And the Knowledge and Cognitive Systems (KCS) program morphed into a program explicitly called Artificial Intelligence and Cognitive Science. As it happens, I ran both the HCI and KCS programs for two years and work closely with all of these programs. However, what I write here is not based on any insider knowledge, but simply on what has been available for any visitor to the NSF website to see. Anybody who wants to know exactly what research all of these programs are funding can search the web-based NSF awards database, which offers abstracts of all grants made during the past 15 years.

Some of the more vocal pioneers of AI claimed that their work was giving machines the power to think the way humans do, but the failure of their technology to match human performance has become a profound embarrassment, now that they can no longer blame the slow speeds and low memory capacity of their computers. Recently, the focus of AI has shifted from supplanting to supplementing human intelligence. The goal is no longer a walking, talking robot, but tools that improve the effectiveness of information systems serving human needs. When the Spirit robot rover was falling toward the Martian surface, immediately before its heroic bounces to a safe land-ing, the onboard computer vision system detected that the spacecraft was moving to one side more rapidly than desired, so the control system immediately compensated. Because there was no human being on board, and the speed-of-light limitation prevented information and commands from going to and from Earth in time, autonomous vision was crucial. But the robot vision system was not capable of recognizing objects, merely measuring the speed that objects and textures were moving across the field of view. Computer vision is progressing rapidly, but it has a long way to go before it matches the capabilities of a sparrow, let alone a human being.

Similarly, the computer techniques called natural language processing (NLP) are progressing rapidly. [23] Imperfect but serviceable speech recognition programs now operate many companies' telephone information and reservation systems, and the quality of this technology improves constantly. Perhaps the greatest controversy in NLP right now is how much can be accomplished by brute-force statistical systems without incorporating the results of linguistics research about the structure of language and the meaning of words. For example, a speech recognition system may consist of a set of mathematical models of phonemes – the individual sounds that comprise speech when strung together – and a statistical model of the probability that various words would appear together in a sentence, based on analysis of a vast corpus of written language using techniques such as Hidden Markov Models (HMM). Some researchers are beginning to add linguistic rules, word definitions, and information from other modes of communication such as facial expressions. But NLP is not yet a fully 'intelligent' system that could take over a person's speaking and listening functions, let alone thought processes.

Artificial Intelligence, broadly defined, covers not only computerized vision and language, but a whole host of other approaches such as probabilistic planning, rule-based reasoning, case-based reasoning, logic programming, machine learning, neural networks [24;25] and even methods that go beyond what individual human minds do, such as genetic algorithms that mimic biological evolution. [26] Perhaps the major application area is finding patterns in vast sets of data. One often hears about knowledge extraction, implying that the computer can take crude data and refine it to meaningfulness. Data fusion and data mining refer to assembling raw data from many sources and sifting them for discoveries that a human could never make, simply because the data are too vast and too fragmented. It is often said that a human being can hold only about seven things in mind at once, whereas modern information systems deal with billions. [27]

Thus, AI intending to duplicate human abilities is currently in hiatus, while AI to enhance the information processing abilities of humans is being pursued aggressively. At the same time, some researchers are doing fundamental research at the intersection of AI and Cognitive Science. For example, Lokendra Shastri [28–30] of the International Computer Science Institute in Berkeley, California, has been modeling how the hippocampal system in the human brain may record memories of specific episodes in the person's life. Damage to the hippocampus deep within the human skull disrupts the ability of the person to learn life's events, without necessarily blocking other kinds of learning such as language and skills. Shastri postulates that the hippocampus and related elements of the brain hold memories of episodes as connections between a small number of concepts that themselves are stored elsewhere. "John gave Mary a book" may be represented by neural connections between a link to memories of John, a link to memories of Mary, a link to the concept book, plus verb links such as that "gives" requires someone in a giver relationship and someone in a receiver relationship.

Nobody knows how many episodes an adult remembers, but the number 50,000 is sometimes mentioned in Cognitive Science discussions. If Shastri is right, then a computer model of the hippocampal system might not need to be very big. If concepts are addressed efficiently in other parts of the brain, then an episode like "John gave Mary a book" could be stored in fewer than a hundred bytes. This would imply the entire hippocampal system could get by with only 5 megabytes of memory, a tiny fraction of the memory of today's pocket computers. Shastri notes that the human mind imaginatively fills in the missing details of memories, and each episodic memory really has very little information in it. We are not generally conscious of the yawning gaps in our memory, any more than we are conscious of the blind spot in the vision of each of our eyes. Shastri also notes that memories are generally stored in multiple copies, perhaps to guard against losing them through the death of any single neuron, but the redundancy in memory may have other functions as well, such as helping us combine facts from different sources by placing some copies of them nearer to each other. Computerized memories might not need that redundancy.

This very quick summary of the current state of AI suggests that we really cannot predict how soon computer and information scientists will be able to simulate real human minds. Rapid progress is going on in other directions, and in a few years a renewed interest in duplicating human intelligence could plausibly move forward very quickly by exploiting all the discoveries and inventions that are being made now for other purposes. Once we know how to duplicate a mind in a computer, then we will know far better how much information of what kinds we will need. We do not currently have this knowledge, but such rapid progress is being made in several approaches to recording human personalities. I would argue that it is time to begin seriously recording people who want it done yet are unlikely to live until the technology is completely mature.

IDENTITY DIFFUSION

In principle, and perhaps in actuality three or four decades from now, it should be possible to transfer a human personality into a robot, thereby extending the person's lifetime by the durability of the machine. This is an old idea that is probably also old-fashioned. A better and more modern idea might be semi-autonomous robots that periodically or continuously update and are updated by a networked database. There is no need to design a vastly expensive, technologically challenging robot into which a human's personality could be placed. Rather, a person archived in a dynamic, distributed information system may temporarily use a variety of relatively simple robots over a period of time, via wireless links. These robots may be modular, reconfigurable, and specialized. There could be aquatic robots for swimming, aerial robots for flying, and mole-like robots for traveling underground - all of which could be shared by many individuals in turn for sake of economy.

One may well ask about a distributed intelligence: Where is it located? We often use traditional language and metaphorically locate ourselves in our hearts, even though that cognition actually takes place in our brains. Subjectively, we are located wherever our senses collect input. Thus, if the hardware that hosts your mind is in a laboratory, but the input and feedback come from an ocean-going robot, then your consciousness is in the sea, not the lab. However, if the robot sinks, your consciousness will revert to the safety of the lab.

Computer archived human personalities will live again on the grid. The grid is more than the net or the web. The Internet is a network of connections that allows one computer to send data to another. The World Wide Web is the most prominent of the many data exchange systems that lie on top of the Internet, consisting of a vast number of data files connected by hypertext links. The Cyberinfrastructure Grid includes the net and the web, but by definition it also includes a variety of physical resources such as sensors and other input devices, actuators and other output devices, memory, and computers. Today, grid computing is chiefly a method of simulating the performance of a supercomputer by linking together a number of more ordinary computers that carry out calculations in parallel. Computer scientists are beginning to imagine a future grid that unites every imaginable resource, which would make it an ideal environment in which to be reincarnated.

Once a personality has entered the grid, it may use a variety of resources located at vast distances from each other, which could require it to separate into a number of fairly autonomous pieces that intermittently communicate or rejoin with each other. No longer contained within the confines of the skull, intelligence will distribute itself dynamically across the information network, becoming potentially ubiquitous. [31] On the one hand, identity diffusion means that the person becomes scattered, possibly incoherent but certainly complex – no longer a star but a nebula. On the other hand, this also means that the person can become greater than any conventional individual, a distributed intelligence that pervades civilization.

CONCLUSION

In the distant future, we may learn to conceptualize our biological lives on Earth as extended childhoods preparing us for the real life that follows in cyberspace. The metaphor of biological caterpillars becoming cybernetic butterflies would be apt, were it not for the proverbial fragility of insects. And the transition from flesh to data will not be so much metamorphosis as liberation. As information contained in a star-spanning database – call it *StarBase* – we will travel across immensity, create new bodies along the way to dwell in every possible environment, and have adventures of the spirit throughout the universe. [32] Fundamentally, we are dynamic patterns of information. The self-awareness that we call consciousness is not a supernatural soul, but the natural consequence of our semantic complexity that gives us the ability to conceptualize ourselves. As information, we can be translated from one storage medium to another, combined with other information, and expressed through an almost infinite variety of instrumentalities. When we emerge into cyberspace, we should no more lament the loss of the bodies that we leave behind than an eagle hatchling laments the shattered fragments of its egg when it first takes wing.

References

1) Roco, MC & Bainbridge, WS; *Societal implications of nanoscience and nanotechnology* (2001); Kluwer

2) Roco, MC & Bainbridge, WS; *Converging technologies for improving human performance* (2003); Kluwer

3) Kurzweil, R; *The Age of Spiritual Machines* (1999); Penguin

4) Bainbridge, WS; "A Question of Immortality" in: *Analog* (2002, Vol. 122–5); pg.40–49

5) Bainbridge, WS; *Survey research: A computer-assisted introduction* (1989); Wadsworth

6) Bainbridge, WS; *Social research methods and statistics: A computer-assisted introduction* (1992); Wadsworth

7) Bainbridge, WS; *The Encyclopedia of Human-Computer Interaction* (2004); Berkshire Publishing

8) Kaplan, B; *Studying Personality Cross-Culturally* (1961); Harper and Row

9) Bainbridge, WS; "Religious Ethnography on the World Wide Web" in: *Religion and the Internet* (2000) edited by Jeffrey K. Hadden and Douglas Cowan; JAI Press

10) Bainbridge, WS; "Validity of Web-Based Surveys," in: *Computing in the Social Sciences and Humanities* (2002) edited by Orville Vernon Burton. University of Illinois Press; pg. 51–66

11) Bainbridge, WS; "The Future of the Internet," in: *Society Online: The Internet in Context* (2004) edited by Philip N. Howard and Steve Jones; Sage; pg. 307–324

12) Bainbridge, WS; "Religion and Science," (2004) forthcoming in *Futures*

13) Wactlar, H & Gong Y; "Informedia Experience-on-Demand: Capturing, Integrating and Communicating Experiences across People, Time, and Space," in: *ACM Computing Surveys* (1999, Vol. 31)

14) Bell, G & Gray J; "Digital Immortality," in: *Communications of the ACM* (2001, Vol. 44); pg. 29–30

15) Gemmell, JG & Bell, & Lueder, R & Drucker, S & Wong C;. "MyLifeBits," in: *ACM Multimedia '02* (2002); New York: Association for Computing Machinery; pg. 235–238

16) Bianchi-Berthouze, N; "Mining Multimedia Subjective Feedback," in: *Journal of Intelligent Information Systems* (2002, Vol. 19); edited by Bianchi-Berthouze; pg. 43–59

17) Thalmann, NMP & Kalra, P & Escher M; "Face to Virtual Face," in: *Proceedings of the IEEE* (1998, Vol. 86); pg. 870–883

18) Lin, W & Hauptman AG; "A Wearable Digital Library of Personal Conversations," in: *Joint Conference on Digital Libraries* (JCDL'02) (2002); pg. 277–278

19) Greenwald, AG & McGhee, DE & Schwartz JLK; "Measuring Individual Differences in Implicit Cognition: The Implicit Association Test" in: *Journal of Personality and Social Psychology* (1998, Vol 74); pg. 1464–1480

20) Greenwald, AG & Farnham SD; "Using the Implicit Association Test to Measure Self-esteem and Self-concept," in: *Journal of Personality and Social Psychology* (2000, Vol. 79); pg. 1022–1038

21) Barrett, LF & Barrett DJ; "An Introduction to Computerized Experience Sampling in Psychology," in: *Social Science Computer Review* (2001, Vol. 19); pg. 175–185

22) Crevier, D; *AI: The Tumultuous History of the Search for Artificial Intelligence* (1993); Basic Books

23) Jurafsky, D & Martin JH; *Speech and Language Processing* (2000); Prentice Hall

24) Bainbridge, WS; 1995a. "Minimum Intelligent Neural Device: A Tool for Social Simulation" in: *Mathematical Sociology* (1995, Vol. 20); pg. 179–192

25) Bainbridge, WS; "Neural Network Models of Religious Belief," in: *Sociological Perspectives* (1995, Vol. 38); pg. 483–

26) Bainbridge, WS; "The Evolution of Semantic Systems," in: *Annals of the New York Academy of Science* (2004)

27) Witten, IH & Moffat, A & Bell TC; *Managing Gigabytes: Compressing and Indexing Documents and Images* (1999); Morgan Kaufmann

28) Shastri, L; 2001. "A Computational Model of Episodic Memory Formation in the Hippocampal System," in: *Neurocomputting* (2001); pg. 889–897

29) Shastri, L; "A Computationally Efficient Abstraction of Long-term Potentiation," in: *Neurocomputing* (2002 Vol. 44–46); pg. 33–41

30) Shastri, L; "Episodic Memory and Cortico-Hippocampal Interactions," in: *Trends in Cognitive Sciences* (2002); pg. 162-168

31) Mitchell, WJ; *ME++: The Cyborg Self and the Networked City* (2003); MIT Press

32) Bainbridge, WS "The Spaceflight Revolution Revisited," in: *Looking Backward, Looking Forward* (2002) edited by Stephen J Garber; Washington, D.C.: National Aeronautics and Space Administration; pg. 39–64

WILL ROBOTS INHERIT THE EARTH?

Marvin L. Minsky, Ph.D.

Early to bed and early to rise, makes a man healthy and wealthy and wise. --- Ben Franklin

Everyone wants wisdom and wealth. Nevertheless, our health often gives out before we achieve them. To lengthen our lives, and improve our minds, in the future we will need to change our bodies and brains. To that end, we first must consider how normal Darwinian evolution brought us to where we are. Then we must imagine ways in which future replacements for worn body parts might solve most problems of failing health. We must then invent strategies to augment our brains and gain greater wisdom. Eventually we will entirely replace our brains – using nanotechnology. Once delivered from the limitations of biology, we will be able to decide the length of our lives – with the option of immortality – and choose among other, unimagined capabilities as well. In such a future, attaining wealth will not be a problem; the trouble will be in controlling it. Obviously, such changes are difficult to envision. Many still argue that these advances are impossible, particularly in the domain of artificial intelligence. But the sciences needed to enact this transition are already in the making, and it is time to consider what this new world will be like.

HEALTH AND LONGEVITY

Such a future cannot be realized through biology. In recent times we have learned a lot about health and how to maintain it. We have devised thousands of specific treatments for particular diseases and disabilities. Scientists are seriously considering the possibility of extending the maximum human life span, but we have not yet achieved this goal. According to the estimates of Roy Walford, professor of pathology at UCLA Medical School, the average human life span was about 22 years in ancient Rome; about 50 in the developed countries in 1900, and today stands at about 75. Still, each of those curves seems to terminate near 115 years [1]. Centuries of improvements in health care have had no effect on that maximum.

Why are our life spans so limited? The answer is simple: Natural selection favors the genes of those with the most descendants. Those numbers tend to grow exponentially with the number of generations. This favors the genes of those who reproduce at earlier ages. Evolution does not usually favor genes that lengthen lives beyond that amount adults need to care for their young. Indeed, it may even favor offspring who do not have to compete with living parents. Such competition could promote the accumulation of genes that cause death. We humans appear to be the longest-lived warm-blooded animals. What selective pressure might have led to our present longevity, almost twice that of our other primate relatives? This is related to wisdom! Among all mammals, our infants are the most poorly equipped to survive by themselves. Perhaps we needed not only parents, but grandparents too, to care for us and to pass on precious survival tips.

Even with such advice, there are many causes of mortality to which we might succumb. Some deaths result from infections. Our immune systems have evolved versatile ways to deal with most diseases. Unhappily though, those very same immune systems often injure us by treating various parts of ourselves as though they too, were infectious invaders. This blindness leads to conditions such as diabetes, multiple sclerosis, rheumatoid arthritis, and many others.

We are also subject to injuries that our bodies cannot repair. Namely, accidents, dietary imbalances, chemical poisons, heat, radiation, and sundry other influences can deform or chemically alter the molecules inside our cells so that they are unable to function. Some of these errors get corrected by replacing defective molecules. However, when the replacement rate is too slow, errors accumulate. For example, when the proteins of the eyes' lenses lose their elasticity, we lose our ability to focus and need bifocal spectacles.

BIOLOGICAL WEARING-OUT

Most likely, eventual senescence is inevitable in all biological organisms. As we learn more about our genes and cellular biochemistry, we will hopefully be able to correct, or at least postpone many conditions that still plague our later years. However, even if we found cures for each specific disease, we would still have to deal with the general problem of 'wearing out'. The normal function of every cell involves thousands of chemical processes, each of which sometimes makes random mistakes. Our bodies use many kinds of correction techniques, each triggered by a specific type of mistake. However, those random errors happen in so many different ways that any low-level scheme to correct them would be difficult indeed.

The problem is that our genetic systems were not designed for very long-term maintenance. The relationship between genes and cells is exceedingly indirect. To repair defects on larger scales, a body would need some sort of catalogue that specified where the various types of cells should be located

– or extensive, constant treatment using future regenerative medicine. In computer programs it is easy to install such redundancy. Many computers maintain unused copies of their most critical 'system' programs, and routinely check their integrity. However, no animals have evolved like schemes, presumably because such algorithms cannot develop through natural selection. The trouble is that error correction then would stop mutation – which would ultimately slow the rate of evolution of an animal's descendants so much that they would be unable to adapt to changes in their environments.

Could we live for several centuries simply by changing some number of genes? After all, we now differ from our evolutionary relatives, the gorillas and chimpanzees, by only a few thousand genes – and yet we live almost twice as long. If we assume that only a small fraction of those new genes caused that increase in life span, then perhaps no more than a hundred or so of those genes were involved. Still, even if this turned out to be true, it would not guarantee that we could gain another century by changing another hundred genes. We might need to change only a few of them – or we might have to change a good many more.

Making new genes and installing them is slowly becoming feasible. But we are already exploiting another approach to combat biological wear and tear: replacing each organ that threatens to fail with a biological or artificial substitute. Some replacements have already become routine, while others are on the horizon. Hearts are merely clever pumps. Muscles and bones are motors and beams. Digestive systems are chemical reactors. Eventually, we will solve the problems associated with transplanting or replacing all of these parts.

When we consider replacing a brain though, a transplant will not work. You cannot simply exchange your brain for another and remain the same person. You would lose the knowledge and the processes that constitute your identity.

Nevertheless, we might be able to replace certain worn out parts of brains by transplanting tissue-cultured stem cells. This procedure would not restore lost knowledge – but that might not matter as much as it seems. We probably store each fragment of knowledge in several different places, in different forms. New parts of the brain could be retrained and reintegrated with the rest, and some of that might even happen spontaneously. Progress in regenerative medicine in the past few years is already leading to this form of treatment for neurodegenerative conditions like Parkinson's.

LIMITATIONS OF HUMAN WISDOM

Even before our bodies wear out, I suspect that we run into limitations of our brains. As a species we seem to have reached a plateau in our intellectual development. There is no sign that we are getting smarter. Was Albert Einstein a better scientist than Newton or Archimedes? Has any playwright in recent years topped Shakespeare or Euripides? We have learned a lot in two thousand years, yet much ancient wisdom still seems sound – which makes me suspect that we haven't been making much progress. We still do not know how to deal with conflicts between individual goals and global interests. We are so bad at making important decisions that, whenever we can, we leave to chance what we are unsure about.

Why is our wisdom so limited? Is it because we do not have the time to learn very much, or that we lack enough capacity? Is it because, as in popular legend, we use only a fraction of our brains? Could better education help? Of course it can, but only to a point. Even our best prodigies learn no more than twice as quickly as the rest. Everything takes us too long to learn because our brains are so terribly slow. It would certainly help to have more time, but longevity is not enough.

The brain, like other finite things, must reach some limits to what it can learn. We do not know what those limits are; perhaps our brains could keep learning for several more centuries. Ultimately, though, we will need to increase their capacity.

The more we learn about our brains, the more ways we will find to improve them. Each brain has hundreds of specialized regions. We know only a little about what each one does – but as soon as we find out how any one part works, researchers will try to devise ways to extend that organ's capacity. They will also conceive of entirely new abilities that biology has never provided. As these inventions accumulate, we will try to connect them to our brains – perhaps through millions of microscopic electrodes inserted into the great nerve-bundle called the corpus callosum, the largest data-bus in the brain. With further advances, no part of the brain will be out of bounds for attaching new accessories. In the end, we will find ways to replace every part of the body and brain – and thus repair all the defects and flaws that make our lives so brief.

REPLACING THE BRAIN

Almost all the knowledge that we learn is embodied in various networks inside our brains. These networks consist of huge numbers of tiny nerve cells, and even larger numbers of smaller structures called synapses, which control how signals jump from one nerve cell to another. To make a replacement of your brain, we would need to know something about how each of your synapses relates to the two cells it bridges. We would also have to know how each of those structures responds to the various electric fields, hormones, neurotransmitters, nutrients and other chemicals that are active in its neighborhood. Your brain contains trillions of synapses, so this is no small requirement.

Fortunately, we would not need to know every minute detail. If that were so, our brains would not work in the first place. In biological organisms, generally each system has evolved to be insensitive to most details of what goes on in the smaller subsystems on which it depends. Therefore, to copy a functional brain, it should suffice to replicate just enough of the function of each part to produce its important effects on other parts.

Suppose that we wanted to copy a machine, such as a brain, that contained a trillion components. Today we could not do such a thing (even were we equipped with the necessary knowledge) if we had to build each component separately. However, if we had a million construction machines that could each build a thousand parts per second, our task would take only minutes. In the decades to come, new fabrication machines will make this possible. Most present-day manufacturing is based on shaping bulk materials. In contrast, the field called 'nanotechnology' aims to build materials and machinery by placing each atom and molecule precisely where we want it.

By such methods, we could make truly identical parts – and thus escape from the randomness that hinders conventionally made machines. Today, for example, when we try to etch very small circuits, the sizes of the wires vary so much that we cannot predict their electrical properties. However, if we can locate each atom exactly, then those wires will be indistinguishable. This would lead to new kinds of materials that current techniques could never make; we could endow them with enormous strength, or novel quantum properties. These products in turn will lead to computers as small as synapses, having unparalleled speed and efficiency.

Limits of Human Memory

If we want to consider augmenting our brains, we might first ask how much a person knows today. Thomas K. Landauer of Bell Communications Research reviewed many experiments in which people were asked to read text, look at pictures, and listen to words, sentences, short passages of music, and nonsense syllables. [1] They were later tested in various ways to see how much they remembered. In none of these situations were people able to learn, and later remember, more than about 2 bits per second, for any extended period. If you could maintain that rate for twelve hours every day for 100 years, the total would be about three billion bits – less than what we can store today on a regular 5-inch Compact Disk. In a decade or so, that amount should fit on a single computer chip.

Although these experiments do not much resemble what we do in real life, we do not have any hard evidence that people can learn more quickly. Despite those popular legends about people with 'photographic memories,' no one seems to have mastered, word for word, the contents of as few as one hundred books – or of a single major encyclopedia. The complete works of Shakespeare come to about 130 million bits. Landauer's limit implies that a person would need at least four years to memorize them. We have no well-founded estimates of how much information we require to perform skills such as painting or skiing, but I do not see any reason why these activities should not be similarly limited.

The brain is believed to contain the order of a hundred trillion synapses – which should leave plenty of room for those few billion bits of reproducible memories. Someday though, it should be feasible to build that much storage space into a package as small as a pea, using nanotechnology.

The Future of Intelligence

Once we know what we need to do, our nanotechnologies should enable us to construct replacement bodies and brains that won't be constrained to work at the crawling pace of 'real time'. The events in our computer chips already happen millions of times faster than those in brain cells. Hence, we could design our new selves to think a million times faster than we do. To such a being, half a minute might seem as long as one of our years, and each hour as long as an entire human lifetime.

But could such beings really exist? Many thinkers firmly maintain that machines will never have thoughts like ours, because no matter how we build them, they will always lack some vital ingredient. They call this essence by various names – like sentience, consciousness, spirit, or soul. Philosophers write entire books to prove that, because of this deficiency, machines can never feel or understand the sorts of things that people do. However, every proof in each of those books is flawed in the same way: by assuming the thing that it purports to prove – the existence of some magical spark that has no detectable properties.

In order to think effectively, you need multiple processes to help you describe, predict, explain, abstract, and plan what your mind should do next. The reason we can think so well is not because we house mysterious spark-like talents and gifts, but because we employ societies of agencies that work in concert to keep us from getting stuck. When we discover how these societies work, we can put them to inside computers too. Then if one procedure in a program gets stuck, another might suggest an alternative approach. If you saw a machine do things like that, you would certainly think it was conscious.

THE FAILURES OF ETHICS

This article bears on our rights to have children, to change our genes, and to die if we so wish. No popular ethical system yet, be it humanist or religion-based, has shown itself able to face the challenges that already confront us. How many people should occupy Earth? What sorts of people should they be? How should we share the available space? Clearly, we must change our ideas about making additional children. Individuals now are conceived by chance. Someday, though, they could be 'composed' in accord with considered desires and designs. Furthermore, when we build new brains, these need not start out the way ours do, with so little knowledge about the world. What sorts of things should our new children know? How many of them should we produce and who should decide their attributes?

Whatever the unknown future may bring, already we are changing the rules that made us. Although most of us will be fearful of change, others will surely want to escape from our present limitations. When I decided to write this article, I tried these ideas out on several groups and had them respond to informal polls. I was amazed to find that at least three quarters of the audience seemed to feel that our life spans were already too long. "Why would anyone want to live for five hundred years? Wouldn't it be boring? What if you outlived all your friends? What would you do with all that time?" they asked. It seemed as though they secretly feared that they did not deserve to live so long. I find it rather worrisome that so many people are resigned to die.

My scientist friends showed few such concerns. "There are countless things that I want to find out, and so many problems I want to solve, that I could use many centuries," they said. Certainly, immortality would seem unattractive if it meant endless infirmity, debility, and dependency upon others – but we will build a state of perfect health. Some people expressed a sounder concern – that the old ones must die because young ones are needed to weed out their worn-out ideas. However, if it is true, as I fear, that we are approaching our intellectual limits, then that response is not a good answer. We would still be cut off from the larger ideas in those oceans of wisdom beyond our grasp. [2]

References

1) Landauer, TK; "How Much Do People Remember? Some Estimates of the Quantity of Learned Information in Long-term Memory", in *Cognitive Science* (1986) pg.10, 477–493

2) This article first appeared in *Scientific American*, October 1994 with some minor revisions.

Medical Time Travel

A question of science

Brian Wowk, Ph.D.

Time travel is a solved problem. Einstein showed that if you travel in a spaceship for months at speeds close to the speed of light, you can return to earth centuries in the future. Unfortunately for would-be time travelers, such spacecraft will not be available until centuries in the future.

Rather than Einstein, nature relies on Arrhenius to achieve time travel. The Arrhenius equation of chemistry describes how chemical reactions slow down as temperature is reduced. Since life is chemistry, life itself slows down at cooler temperatures. Hibernating animals use this principle to time travel from summer to summer, skipping winters when food is scarce.

Medicine already uses this kind of biological time travel. When transplantable organs such as hearts or kidneys are removed from donors, the organs begin dying as soon as their blood supply stops. Removed organs have only minutes to live. However with special preservation solutions and cooling in ice, organs can be moved across hours of time and thousands of miles to waiting recipients. Cold slows chemical processes that would otherwise be quickly fatal.

Can whole people travel through time like preserved organs? Remarkably, the answer seems to be yes. Although it is seldom done, medicine sometimes does preserve people like organs awaiting transplant. Some surgeries on major blood vessels of the heart or brain can only be done if blood circulation through the entire body is stopped. [1;2] Stopped blood circulation would ordinarily be fatal within 5 minutes, but cooling to +16°C (60°F) allows the human body to remain alive in a 'turned off' state for up to 60 minutes. [3] With special blood substitutes and further cooling to a temperature of 0°C (32°F), life without heartbeat or circulation can be extended as much as three hours. [4] Although there is currently no surgical use for circulatory arrest of several hours [5], it may be used in the future to permit surgical repair of wounds before blood circulation is restored after severe trauma. [6]

While some biological processes are merely slowed by deep cooling, others are completely stopped. Brain activity is an important example. Brain electrical activity usually ceases at temperatures below +18°C (64°F), and disappears completely in all cases as freezing temperatures are approached. [7] Yet these temperatures can still be survived. In fact, not only can the brain survive being turned off, surgeons often use drugs to force the brain to turn off when temperature alone does not do the trick. [8] They do this because if the brain is active when blood circulation is stopped, vital energy stores can become depleted, later causing death. This reminds us that death is not when life turns off. Death is when the chemistry of life becomes irreversibly damaged.

Specialized surgeries are not the only cases in which the brain can stop working and later start again. Simple cardiac arrest (stopping of the heart) at normal body temperature also causes brain electrical activity to stop within 40 seconds. [9] Yet the heart can remain stopped for several times this long with no lasting harm to the brain. Anesthetic drugs, such as barbiturates, can flatten EEG (brain electrical activity) readings for many hours while still permitting later recovery. [10] This prolonged drug-induced elimination of brain activity is sometimes used as a treatment for head injuries. [11] Patients do not emerge from these comas as blank slates. Evidently human beings do not require continuous operation like computer chips. Brains store long-term memories in physical structures, not fleeting electrical patterns.

Perhaps the most extreme example of brains completely stopping and later starting again are the experiments of Isamu Suda reported in the journal Nature [12] and elsewhere [13] in 1966 and 1974. Suda showed recovery of EEG activity in cat brains resuscitated with warm blood after frozen storage at -20°C (-4°F) for up to seven years.

Reversible experiments in which all electrical activity stops, and chemistry comes to a virtual halt, disprove the 19th-century belief that there is a 'spark of life' inside living things. Life is chemistry. When the chemistry of life is adequately preserved, so is life. When the chemical structure and organization of a human mind is adequately preserved, so is the person.

Suda's frozen cat brains deteriorated with time. Brains thawed after five days showed EEG patterns almost identical to EEGs obtained before freezing. However brains thawed after seven years showed greatly slowed activity. At a temperature of -20°C, liquid water still exists in a concentrated solution between ice crystals. Chemical deterioration still slowly occurs in this cold liquid.

Preserving the chemistry of life for unlimited periods of time requires cooling below -130°C (-200°F). [14] Below this temperature, any remaining unfrozen liquid between ice crystals undergoes a "glass transition". Molecules become stuck to their neighbors with weak hydrogen bonds. Instead of wandering about, molecules just vibrate in one place. Without freely moving molecules, all chemistry stops.

For living cells to survive this process, chemicals called cryo-protectants must be added. Cryoprotectants, such as glycerol, are small molecules that freely penetrate inside cells and limit the percentage of water that converts into ice during cooling. This allows cells to survive freezing by remaining in isolated pockets of unfrozen solution between ice crystals. [14] Below the glass transition temperature, molecules inside these pockets lock into place, and cells remain preserved inside the glassy water-cryoprotectant mixture between ice crystals.

This approach for preserving individual cells by freezing was first demonstrated half a century ago. [15] It is now used routinely for many different cell types, including human embryos. Preserving organized tissue by freezing has proven to be more difficult. While isolated cells can accommodate as much as 80% of the water around them turning into ice, organs are much less forgiving because there is no room between cells for ice to grow. [16] Suda's cat brains survived freezing because the relatively warm temperature of -20°C allowed modest quantities of glycerol to keep ice formation between cells within tolerable limits.

In 1984 cryobiologist Greg Fahy proposed a new approach to the problem of complex tissue preservation at low temperature. [17] Instead of freezing, Fahy proposed loading tissue with so much cryoprotectant that ice formation would be completely prevented at all temperatures. Below the glass transition temperature, entire organs would become a glassy solid (a solid with the molecular structure of a liquid), free of any damage from ice. This process was called "vitrification". Preservation by vitrification, first demonstrated for embryos [18], has now been successfully applied to many different cell types and tissues of increasing complexity. In 2000, reversible vitrification of transplantable blood vessels was demonstrated. [19]

New breakthroughs in reducing the toxicity of vitrification solutions [20], and in adding synthetic ice blocking
molecules [21;22] continue to push the field forward.
In 2004, successful transplantation of rabbit kidneys after
cooling to a temperature of -45°C (-49°F) was reported.
[23] These kidneys were prevented from freezing by replacing more than half of the water inside them with vitrification
chemicals. Amazingly, organs can survive this extreme treatment if the chemicals are introduced and removed quickly at
low temperature.

Reversible vitrification of major organs is a reasonable prospect within this decade. What about vitrification of whole
animals? This is a much more difficult problem. Some organs,
such as the kidney and brain, are privileged organs for vitrification because of their high blood flow rate. This allows
vitrification chemicals to enter and leave them quickly before
there are toxic effects. Most other tissues would not survive
the long chemical exposure time required to absorb a sufficient concentration to prevent freezing.

It is useful to distinguish between reversible vitrification and
morphological vitrification. Reversible vitrification is vitrification in which tissue recovers from the vitrification process
in a viable state. Morphological vitrification is vitrification in
which tissue is preserved without freezing, with good structural
preservation, but in which key enzymes or other biomolecules
are damaged by the vitrification chemicals. Morphological
vitrification of a kidney was photographically demonstrated
in Fahy's original vitrification paper [17], but 20 years later
reversible kidney vitrification is still being pursued.

Given this background, what are the prospects of reversibly
vitrifying a whole human being? It is theoretically possible,
but the prospects are still distant. Morphological vitrification
of most organs and tissues in the body may now be possible,
but moving from morphological vitrification to reversible

vitrification will require fundamental new knowledge of mechanisms of cryoprotectant toxicity, and means to intervene in those mechanisms.

If reversible vitrification of humans is developed in future decades, what would be the application of this 'suspended animation'? Space travel is sometimes suggested as an application, but time travel – specifically, medical time travel – seems more likely to be the primary application. People, especially young people dying of diseases expected to be treatable in future years would be most motivated to try new suspended animation technologies. Governments would probably not even allow anyone but dying people to undergo such an extreme process, especially in the early days. Applications like space travel would come much later.

Medical time travel, by definition, involves technological anticipation. Sometimes this anticipation goes beyond just cures for disease. After all, if people are cryopreserved in anticipation of future cures, what about future cures for imperfections of the preservation process itself? As the medical prospect of reversible suspended animation draws nearer, the temptation to cut this corner will become stronger. In fact, some people are already cutting this corner very wide.

In 1964, with the science of cryobiology still in its infancy, Robert Ettinger proposed freezing recently deceased persons until science could resuscitate them. [24] The proposal assumed that the cause of death, the early stages of clinical death, and crude preservation would all be reversible in the future. Even aging was to be reversed. This proposal was made in absence of any detailed knowledge of the effects of stopped blood flow or freezing on the human body. The proposal later came to be known as 'cryonics'.

Cryonics was clever in that it circumvented legal obstacles to cryopreserving people by operating on the other side of the legal dividing line of death. However 40 years later, as measured by the number of people involved and the scientific acceptance of the field, cryonics remains a fringe practice. Why? Probably because by operating as it does, cryonics is perceived as interment rather than medicine. One organization, the Cryonics Institute, is even licensed as a cemetery. It advertises that professional morticians deliver its services (as if this is an endorsement?). Dictionaries now define cryonics as 'freezing a dead human'. Is it any wonder that cryonics is unpopular? It is a failure by definition!

Is this view biologically justified? In the 1980's another cryonics organization, the Alcor Life Extension Foundation, adopted a different approach to cryonics. Under the leadership of cardiothoracic surgery researcher, Jerry Leaf, and dialysis technician, Mike Darwin, Alcor brought methods of modern medicine into cryonics. Alcor sought to validate each step of their cryopreservation process as reversible, beginning with life support provided immediately after cardiac arrest, and continuing through hours of circulation with blood replacement solutions. Leaf and Darwin showed that large animals could be successfully recovered after several hours at near-freezing temperatures under conditions similar to those in the first hours of real cryonics cases. [25] Blood gas measurements and clinical chemistries obtained in real cryonics cases further demonstrated that application of life support techniques (mechanical CPR and heart-lung machines) could keep cryonics subjects biologically alive even in a state of cardiac arrest and legal death. [26]

This leaves cryonics today in an interesting situation. It is stigmatized as something that cannot work because the subjects are legally deceased. Yet under ideal conditions the subjects are apparently alive by all measurable criteria, except heartbeat. At the start they are biologically the same as patients undergoing open heart surgery, legal labels notwithstanding. The cryopreservation phase of cryonics is of course not yet reversible. But cryonicists would argue that this does not imply death either because death only happens when biochemistry becomes irreversibly damaged, and "irreversibility" is technology-dependent.

To clarify these issues, cryonicists have proposed the "information-theoretic criterion" for death. [27] According to this criterion, you are not dead when life stops (we already know that from clinical medicine), you are not dead when biochemistry is damaged, you are only dead when biochemistry is so badly damaged that no technology, not even molecular nanotechnology [28], could restore normal biochemistry with your memories intact. By this criterion, someone who suffered cardiac arrest days ago in the wilderness is really dead. Someone who suffered only a few minutes of cardiac arrest and cryoprotectant toxicity during morphological vitrification may not be.

Whether or not one accepts this information-theoretic criterion, the modern cryonics practice of using life support equipment to resuscitate the brain after legal death raises important issues. Among them is the scientific issue that cryonics cannot be dismissed simply by calling its subjects 'dead'. Two minutes of cardiac arrest followed by restoration of blood circulation does not a skeleton make. There should be a rule that no one is allowed to say "dead" when discussing cryonics. It is usually a slur that communicates nothing scientific.

Whether cryonics can work depends on biological details of cerebral ischemic injury (brain injury during stopped blood flow), cryopreservation injury, and anticipated future technology. There is much published literature on cerebral ischemia, and a small, but growing body of writing on relevant future technologies. [29– 33] There is, however, very little information on the quality of preservation achieved with cryonics. [34;35] It would seem logical to look to cryobiologists for this information.

Cryobiologists, professional scientists who study the effect of cold on living things, decided long ago that they did not want their field associated with cryonics. [36] The Society for Cryobiology bylaws even provide for the expulsion of members that practice or promote "freezing deceased persons". The result has been the polarization of cryobiologists into either outspoken contempt or silence concerning cryonics. The contempt camp typically speaks of cryonics as if it has not changed in 40 years. This political environment, plus the fact that most cryobiologists work outside the specialty of organ cryopreservation, makes obtaining informed cryobiological information about cryonics very difficult.

The most important cryobiological fact of cryonics (other than its current irreversibility) is that cryoprotectant chemicals can be successfully circulated through most of the major organs of the body if blood clots are not present. We can conclude this by simply considering that everything now known about long-term preservation of individual organs was learned by removing and treating those organs under conditions similar to ideal cryonics cases. It is generally observed that the quality of cell structure preservation (as revealed by light and electron microscopy) is very poor when there is no cryoprotectant, but steadily improves as the concentration of cryoprotectant is increased (provided toxicity thresholds are not exceeded). Recent years have seen a trend toward using higher cryoprotectant concentrations in cryonics, yielding structural preservation that is impressively similar to unfrozen tissue. [35]

Extraordinary claims require extraordinary evidence. While it is plausible that mechanical CPR and rapid cooling can keep the brain alive long after cardiac arrest, cryonicists need to provide more oxygenation and blood chemistry data from more cases to support this claim. While it is plausible that morphological vitrification of major organs can now be achieved with existing technology, more research needs to be published supporting this claim. Copious technical information is critical to the evaluation of a highly speculative field that will not have clinical feedback for decades or centuries. Cryonics without feedback is a recipe for mischief. [37]

Somewhere between freezing, morphological vitrification, reversible vitrification of the central nervous system, and reversible vitrification of whole people, there is technology that will lead medicine to take the idea of medical time travel seriously within this century. Whether what is now called cryonics will eventually become that technology remains to be seen. It will depend on whether cryonicists can manage to outgrow the stigma attached to their field, and develop methods that are validated by more biological feedback and less hand waving. It may also depend on whether critics of cryonics can manage to engage in more substantive discussion and less name-calling. The ultimate feasibility of medical time travel is a question of science, not rhetoric.

References

1) Aebert H & Brawanski A & Philipp A & Behr R & Ullrich OW & Keyl C & Birnbaum DE; "Deep hypothermia and circulatory arrest for surgery of complex intracranial aneurysms" in: *European Journal of Cardiothoracic Surgery* (1998, Vol. 13); pg. 223–229

2) Ehrlich M & Grabenwoger M & Simon P & Laufer G & Wolner E & Havel M; "Surgical treatment of type A aortic dissections. Results with profound hypothermia and circulatory arrest" in: *Texas Heart Institute Journal* (1995, Vol. 22); pg. 250–253

3) Rosenthal E; "Suspended Animation – Surgery's Frontier" in: *New York Times* (1990, Nov. 13)

4) Haneda K & Thomas R & Sands MP & Breazeale DG & Dillard DH; "Whole body protection during three hours of total circulatory arrest: an experimental study" in: *Cryobiology* (1986, Vol. 23); pg. 483494

5) Greenberg MS; "General technical considerations of aneurysm surgery" in: *Handbook of Neurosurgery* (1997, 4th edition)

6) Bellamy R & Safar P & Tisherman SA & Basford R & Bruttig SP & Capone A & Dubick MA & Ernster L & Hattler BG Jr & Hochachka P & Klain M & Kochanek PM & Kofke WA & Lancaster JR & McGowan FX Jr & Oeltgen PR & Severinghaus JW & Taylor MJ & Zar H; "Suspended animation for delayed resuscitation" in: *Critical Care Medicine* (1996, Vol. 24); pg. S2447

7) Stecker MM & Cheung AT & Pochettino A & Kent GP & Patterson T & Weiss SJ & Bavaria JE; "Deep hypothermic circulatory arrest: I. Effects of cooling on electroencephalogram and evoked potentials" in: *Annals of Thoracic Surgery* (2001, Vol. 71); pg. 14–21

8) Rung GW & Wickey GS & Myers JL & Salus JE & Hensley FA Jr & Martin DE; "Thiopental as an adjunct to hypothermia for EEG suppression in infants prior to circulatory arrest" in: *Journal of Cardiothoracic and Vascular Anesthesia* (1991, Vol. 5); pg. 337–342

9) Lind B & Snyder J & Kampschulte S & Safar P; "A review of total brain ischaemia models in dogs and original experiments on clamping the aorta" in: *Resuscitation* (1975, Vol. 4); pg. 19–31

10) Bird TD & Plum F; Recovery from barbiturate overdose coma with a prolonged isoelectric electroencephalogram" in: *Neurology* (1968, Vol. 18); pg. 456–460

11) Toyama T, in: *Barbiturate Coma*, http://www.trauma.org/anaesthesia/barbcoma.html

12) Suda I & Kito K & Adachi C; "Viability of long term frozen cat brain in vitro" in: *Nature* (1966, Vol. 212); pg. 268–270

13) Suda I & Kito K & Adachi C; "Bioelectric discharges of isolated cat brain after revival from years of frozen storage" in: *Brain Research* (1974, Vol. 70); pg. 527–531

14) Mazur P; "Freezing of living cells: mechanisms and implications" in: *American Journal of Physiology* (1984, Vol. 247); pg. C125–142

15) Polge C & Smith A & Parkes AS; "Revival of Spermatozoa after Vitrification and Dehydration at Low Temperatures" in: *Nature* (1949, Vol. 164); pg. 666

16) Fahy GM & Levy DI & Ali SE; "Some emerging principles underlying the physical properties, biological actions, and utility of vitrification solutions" in: *Cryobiology* (1987, Vol. 24); pg. 196–213

17) Fahy GM & MacFarlane DR & Angell CA & Meryman HT; "Vitrification as an approach to cryopreservation" in: *Cryobiology* (1984, Vol. 21); pg. 407–426

18) Rall WF & Fahy GM; "Ice-free cryopreservation of mouse embryos at -196 degrees C by vitrification" in: *Nature* (1985, Vol. 313); pg. 573–575

19) Song YC & Khirabadi BS & Lightfoot F & Brockbank KG & Taylor MJ; "Vitreous cryopreservation maintains the function of vascular grafts" in: *Nature Biotechnology* (2000, Vol. 18); pg. 296–299

20) Fahy GM & Wowk B & Wu J & Paynter S; "Improved vitrification solutions based on the predictability of vitrification solution toxicity" in: *Cryobiology* (in press)

21) Wowk B & Leitl E & Rasch CM & Mesbah-Karimi N & Harris SB & Fahy GM; Vitrification enhancement by synthetic ice blocking agents" in: *Cryobiology* (2000, Vol. 40); pg. 228–236

22) Wowk B & Fahy GM; "Inhibition of bacterial ice nucleation by polyglycerol polymers" in: *Cryobiology* (2002, Vol. 44); pg. 1423

23) Fahy GM & Wowk B & Wu J & Phan J & Rasch C & Chang A & Zendejas E; "Cryopreservation of Organs by Vitrification: Perspectives and Recent Advances" in: *Cryobiology* (in press)

24) Ettinger RCW; *The Prospect of Immortality* (1964, 1st edition); Doubleday & Company

25) Alcor Life Extension Foundation website: Alcor's Pioneering Total Body Washout Experiments, http://www.alcor.org/Library/html/tbw.html

26) Darwin M; "Cryopreservation of CryoCare Patient #C-2150" in: *Biopreservation Technique Briefs* (1996, Vol 18); http://www.cryocare.org/index.cgi?subdir=bpi&url=tech18b.txt

27) Merkle RC; "The technical feasibility of cryonics" in: *Medical Hypotheses* (1992, Vol. 39); pg. 6–16

28) Drexler E; *Engines of Creation* (1986, 1st edition); Anchor Press/Doubleday

29) Darwin M; "The Anabolocyte: A Biological Approach to Repairing Cryoinjury" in: *Life Extension Magazine* (1977, July/August); pg. 80–63

30) Drexler KE; "Molecular engineering: An approach to the development of general capabilities for molecular manipulation" in: *Proceedings of the National Academy of Sciences* (1981, Vol. 78); pg. 5275–5278

31) Donaldson T; "24th Century Medicine" in: *Analog ScienceFiction/Science-Fact* (1988, Sept.); http://www.alcor.org/Library/html/24thcenturymedicine.html

32) Freitas RA; *Nanomedicine, Vol. I: Basic Capabilities* (1999, 1st edition); Landes Bioscience

33) Freitas RA; *Nanomedicine, Vol. IIA: Biocompatibility* (2003, 1st edition), Landes Bioscience

34) Alcor staff; "Histological study of a temporarily cryopreserved human" in: *Cryonics* (1984, November); pg. 13–32 http://www.alcor.org/Library/html/HistologicalStudy.htm

35) Darwin M & Russell R & Wood L & Wood C; "Effect of Human Cryopreservation Protocol on the Ultrastructure of the Canine Brain" in: *Biopreservation Tech Briefs* (1995, Vol. 16) http://www.cryocare.org/index.cgi?subdir=bpi&url =tech16.txt

36) Darwin M; "Cold War: The Conflict Between Cryonicists and Cryobiologists" http://www.alcor.org/ Library/html/coldwar.html

37) Darwin M; "The Myth of the Golden Scalpel" in: *Cryonics* (1986, January); pg. 15–18

CHAPTER II: PERSPECTIVES

Ethics, Sociology and Philosophy

We could end it here. The scientific story has been told, the experts have made their predictions, and the options have been presented. But the Institute's mission has always been more encompassing. Many questions have been brought up: About what it means to be human, about what it means to be mortal; about the society of the future and the dreams that shape it today. In this section, we will encounter those who are enthusiastically supportive and those who are deeply skeptical of the quest for immortality.

But this section is not just about moral wrongs and (human) rights. We are also asked to consider deeper philosophical questions about time, identity, and our outlook on death and life.

We begin with "**Some Ethical and Theological Considerations**" by Brad F. Mellon. The editors must confess that in light of the recent statements made by the US President's Council on Bioethics, we were pleasantly surprised to encounter such a measured and thoughtful analysis of the relationship between Christianity and the scientific conquest of death. In concluding, Mellon leaves us with at least two questions: Why should we fear death and should we spend resources more wisely?

The latter question is often paraphrased as a Malthusian concern about limited resources. Surely there are too many people already? Yet, immortalist philosopher and founder of the extropian transhumanist movement, **Max More,** argues that "**Superlongevity without Overpopulation**" is entirely feasible.

Another instinctive objection to the scientific conquest of death is to claim that dying is, after all, natural. Businessman and activist **Mike Treder** takes issue with the contention that this is "**Upsetting the Natural Order.**" He sees death as an evil to be eradicated, and the desire for immortality to be far from unnatural – as do many of our scientific contributors.

Eric S. Rabkin, Professor of English Language, examines the way in which the human struggle for immortality has been represented in literature. In a thorough and insightful investigation he comes to conclude that the desire for immortality is "**The Self-defeating Fantasy**". Opposing the preceding author, who advocates the expansion of consciousness by merging digital selves into 'super-beings,' Rabkin warns, "Who would choose such a neutered eternity?"

We can see that there is another dimension in the discussion of life span: identity and its conception. **Dr. Manfred Clynes** leads us in a challenging discussion on "**Timeconciousness in Very Long Life**". If the time we experience is more important than the length of time we live, how would it alter our identity if we were conscious of time in a different way?

After such abstract excursions, some readers will no doubt be pleased to come upon an essay by a true 'identity' who is by no means "neutered": **Shannon Vyff**, mother of three, is a real life immortality advocate who practices caloric restriction, is signed up for cryonic suspension and lobbies for life extension research in her spare time. In her "**Confessions of a Proselytizing Immortalist**" she shares her own story, thoughts and experiences.

But should someone like Shannon really call herself an 'Immortalist?' **Ben Best**, President of the Cryonics Institute, himself a firm advocate for conquering death, feels there are

"Some Problems with Immortalism." Immortality is an inconceivably long time, after all. Should those wishing to conquer death not just focus on extending the human life span?

"On the contrary," replies **Marc Geddes**. In his **"Introduction to Immortalist Morality"** he develops an argument from moral philosophy, grounding moral theory on the human perception of death and the desire for immortality. Geddes also debunks the commonly held notion that our mortality is what makes life worth living.

This leads us to the last essay in this section, which returns to the first question raised by Chaplain Mellon: All this talk of scientific immortality notwithstanding, why **"Should We Fear Death?"** Australian writer **Russell Blackford** examines Epicurean and modern arguments concerning this issue. His statement "We should not console ourselves with false reassurances about the supposed virtues of being mortal" brings a conclusion to this second section.

Some Ethical and Theological Considerations

Brad F. Mellon, Ph.D.

The Immortality Institute (hereafter 'the Institute') is dedicated to the goal of achieving physical immortality through its stated mission to overcome involuntary death. [1] This paper will explore some of the ethical and theological considerations that, in my view, need to be clearly understood in undertaking such an ambitious project. When exploring ethical concerns, I will appeal to the classic Georgetown principles of modern bioethics, namely autonomy, beneficence, nonmaleficence and justice. I will also consult more specific formulations such as the four interests of the Commonwealth of Pennsylvania (to preserve life, prevent suicide, protect third parties, and uphold the integrity of health care facilities). Theological considerations for the present study are taken from the Judeo-Christian tradition, including the Hebrew and Christian Scriptures and theological reflection by noted scholars. Finally, we will need to take into consideration Delkeskamp-Hayes' correct observation that ethics can be viewed as an 'essential ingredient' of theology.[2] The result is that ethical implications are often included or embedded in theological concerns, and should be applied together, not separately.

POSITIVE ASPECTS

In light of ethical and theological principles, the concept of human physical immortality has much to commend it. First, there is an abundance of Scriptures that uphold and promote life, including eternal life. According to Ecclesiastes 3:11, for example, God has placed immortality in our hearts (although we are not able to comprehend it). The wisdom of Proverbs 12:28 contends that the 'path' of the godly leads to life eternal. In Genesis 9:1–6, the sanctity of human life is an integral part of God's covenant agreement with Noah, where the ancient writer connects the sacred character of life to our creation in the Imago Dei (the image of God).

Another passage from the Hebrew Scriptures, Psalm 139:13–16 provides a beautifully poetic description of life as the creative activity of God:

> For you created my inmost being; you knit me together in my mother's womb. I praise you because I am fearfully and wonderfully made; your works are wonderful, I know that full well. My frame was not hidden from you when I was made in the secret place. When I was woven together in the depths of the earth, your eyes saw my unformed body. All the days ordained for me were written in your book before one of them came to be.

Further Scriptures that support the Institute's commitment to radical life extension include Psalm 116, where the author gives thanks to God for delivering him from death and allowing him to live a while longer.

The Christian Scriptures likewise are filled with words that uphold life. For example, Jesus strongly affirms that human life is far more valuable than the resources that are required to sustain it (Matthew 6:25). The reflections of Pope John

Paul II in his treatise entitled, *The Gospel of Life* resound with those of Jesus and deserve to be quoted in full:

> The Gospel of life is at the heart of Jesus' message. Lovingly received day after day by the Church, it is to be preached with dauntless fidelity as 'good news' to the people of every age and culture.... When he presents the heart of his redemptive mission, Jesus says: 'I came that they may have life, and have it abundantly' (John 10:10). In truth, he is referring to that 'new' and 'eternal' life which consists in communion with the Father, to which every person is freely called in the Son by the power of the Sanctifying Spirit. It is precisely in this 'life' that all aspects and stages of human life achieve their full significance. [3]

He goes on to say that God himself has placed "inestimable value" upon our temporal life here on earth. Life is a "sacred reality" that has been entrusted to us, and the result is a responsibility to preserve our own life and that of others.

The ambitious mission of the Institute to eradicate what its members see as the "blight" of involuntary death can also be seen as consistent with at least two of the interests set forth by the Commonwealth of Pennsylvania. In an apparent attempt to assure that every effort will be made to prevent premature death, the Commonwealth has placed a premium on preserving life and protecting third parties. 'Third parties' are those who are financially dependent on their parents, other family members, or guardians. It is not hard to imagine how efforts to preserve and extend physical life might benefit such dependents.

Another positive aspect of the Institute's mission is that it can serve as a counterbalance to what many would call the present-day 'culture of death'. [4] Peter Singer, for example, has expounded the view that human beings should not be set above other forms of life and gives the illustration that an animal, such as a dog or pig might have superior rationality to a severely handicapped human infant. He adds that it might well be kinder to such an infant to offer treatment that leads to death. Singer contends this matter ultimately should be left up to the wishes of the parents. In his view we should respect the desire of a rational human who wants to die and asserts that to give a lethal injection may in certain cases (such as persistent vegetative state) be ethically equivalent to removing a feeding tube (and in his mind preferable and more merciful). [5]

There are other proponents in favor of active euthanasia and assisted suicide for those who suffer. As a representative of this view Jack Kevorkian suggests such measures are "long overdue" in our society. He contends that Western culture has established 'arbitrary laws' against euthanasia and assisted suicide because of pressure brought on by religious beliefs. [6]

Although advocates of euthanasia and assisted suicide contend that such practices are designed in theory to 'do good' (beneficence); many critics claim that the application of these measures can lead in a different and even dangerous direction. An editorial in *Christianity Today*, for example, analyzes recent legislation in Holland that extends a right to die for anyone 16 years old without parental consent. The editor writes:

> In Germany, the moral memory of Aktion T4, Hitler's euthanasia law, is still alive. But the Dutch seem to have forgotten that Hitler's regime first sharpened its execution skills and tested its gas chambers on sick children and disabled adults from 1939 to 1941 before it applied its new technical expertise to Jews at Auschwitz and Treblinka. [7]

One can see why in light of historical developments, 'preventing suicide' is included in various ethical formulations and why Pennsylvania, for example, has established it as one of four state interests. The movement in favor of a 'right to die' in our view needs to be assessed against the movement to radically extend life. At the very least it is our hope that the Institute's pronounced emphasis on life would serve to counter any moves to allow autonomous premature death, or especially the unlawful taking of life (nonmaleficence). Although the Institute's mission statement is directed toward ending involuntary death, adding a position on voluntary death in our view might serve to provide a more realistic perspective on crucial life and death issues facing society.

Concerns

Although we have found support for the Institute's mission among the ethical and theological principles derived from our Judeo-Christian tradition, some concerns arise.

First, there is the obvious reality of death. An American proverb contends that 'death and taxes' are the only two certainties we can expect. The Hebrew and Christian Scriptures likewise testify to death's reality. If we revisit the Scriptures that promote life (above), we find many of them mention death as well. The testimony about immortality within the human soul in Ecclesiastes 3:11 is tempered by a later passage where the writer describes the aging process that brings us to life's end (Ecclesiastes 12:1–7). Shortly after Jesus' affirmation of life in John 10:10 (see above), he clearly indicates that eternal life exists beyond this one (John 14:1–4). Pope John Paul II affirms the profound meaning of life on earth, yet says that human life far exceeds the temporal plane because it is bound up in the very life of God. [3] Drane reminds us that although life, from a Christian perspective, is considered a "gift from God, a creation in the image of God, an object of divine providence", so death also is ordained of God. Even the time of death and the way in which someone dies are part of the divine order. [8]

Death is portrayed in the Scriptures in both a negative and positive light and as a boundary between the human and the Divine. Clowney's vivid description of death demonstrates that it is not something to be taken lightly:

> The brevity of man's life [stands] in fearful contrast with God's eternity.... Death's shadow flies upon us and blots out today's sunlight with tomorrow's darkness. [9]

Death reminds us that we depend upon God for our existence (cf. Acts 17:28), and according to Barth it forms a limit between God and humankind. [10] Although the author of Psalm 116 begins with a thanksgiving to God for extending his life, he later declares that the death of God's people can be "precious" in the Lord's sight.

These Scriptures lead us to the end-of-life bioethical concept of a 'good death' (the original and literal meaning of 'euthanasia'). The authors of a book entitled *Dying Well* describe a good death in ideal terms,

> ...ending one's days in old age, relieved of disabling pain, surrounded by friends and family, attended by sensitive caregivers, reconciled with all persons...at peace with God. [11]

Of course, there are many other dynamics to be considered, such as when death comes in a violent way to the young, etc. However, if a 'good death' is possible, and if eternal life is understood as an existence beyond this one, we are faced with the question of what value can be derived from extending physical life and from efforts to eradicate death.

Finally we will consider comments and analysis by Daniel Callahan and James Drane that serve to further challenge the drive to conquer death. [12] Callahan notes that although "death is treated as an evil in and of itself with no redeeming features (unless, now and then, as a surcease from pain)", and that this war is treated as an imperative, it is nonetheless a relatively modern concept (since Descartes and Bacon). He adds there are problems with this war, such as when a terminally ill patient extends his or her suffering by coming to hospice late in the dying process. Technological advances can also be as much bane as blessing when it is assumed that "something more can always be done" for dying patients. He further suggests that too often in this scenario medical staff unfortunately can ignore a patient's last wishes.

Callahan contends "to fear and resist death might be a perfectly sensible response except for the fact that it fails to ask the question of the *meaning* (italics mine) of death." Likewise it does not adequately deal with quality of life issues. For example, Callahan cannot accept the idea that extending life could offer a guarantee of indefinite freedom from boredom and other problems associated with the aging process. Drane reminds us that another common problem associated with aging is depression, and that extending life and waging war on disease have not solved the problem of lack of meaning for the elderly. Further, he says that ignoring death in older age tends to exacerbate one's problems. Even if we might be able to conveniently ignore death for a time, it can come suddenly and unexpectedly. [8]

Both Callahan and Drane agree that despite efforts to eradicate involuntary death, death will have the final word. This conclusion is one that is consistent with Judeo-Christian theology and ethics. Callahan contends that despite victories in the war on death, people will continue to die. Drane's comments on this subject are likewise pointed:

> At the fringes of every aging experience is increasing pressure from the reality of death. Many of the senior activities in American culture come over as distractions from, or even denials of this reality. Death in the U.S. often is treated as a taboo topic. Sooner or later however, death and questions about how to die, force themselves into consideration. Aging anticipates something else, and that something else is death. Death is a part of the aging experience that cannot be ignored, no matter what the cultural peculiarities of the newly designed period. [8]

Alongside the reality of death are questions directed to how death might theoretically be eradicated. These questions include whether death is conquerable because the time of death is not 'fixed' and comes to different persons at different times. [12] Another suggestion is to see death as a series of potentially preventable diseases that science could conquer by eliminating one disease at a time. [13] In light of the overwhelming historical evidence regarding the reality of death, however, neither of these theories is convincing.

Conclusion and Proposals

The goal of eradicating involuntary death is both supported and challenged by Judeo-Christian theology and ethical principles based on that theology. The Scriptures uphold and promote life, including eternal life, yet view immortality as an existence that goes beyond this temporal, earthly one. Modern ethical formulations that issue a call to preserve life likewise recognize the reality of death. Quality of life issues, such as the quest for meaning in old age and the problems of boredom and depression need to be addressed amid the quest to conquer death and radically extend life.

Callahan has pointed out problems with viewing death as an 'evil' and with committing resources and energy toward a 'war' on death. One of his greatest concerns is how this war produces casualties, especially terminally ill patients who extend their suffering by seeking hospice late in the dying process. Drane raised the issue that ignoring death can become problematic for elderly persons who have to face it suddenly and unexpectedly.

Based on the above discussion, the present writer would like to offer some proposals for the Institute to consider: First, the ambitious nature of the mission statement suggests a need to devote further research to the subject of how death might be eliminated. New theories can be formulated and explored. Second, the mission could be extended to include voluntary as well as involuntary death, which would be compatible with theological and ethical proscriptions against premature or unlawful death, including suicide, assisted suicide, euthanasia, etc. Third, since it would appear that an effort to radically extend life is theoretically more attainable than conquering death; why not commit the greater share of time, energy, and resources to that end? Fourth, considering quality of life issues that emerge during the aging process, we would suggest the Institute seek to address how such problems might be mitigated or resolved.

References

1) Immortality Institute 2003: Constitution & Bylaws, http:// imminst.org/about/constitution.php

2) Delkeskamp-Hayes, Corrina; ""The Price of Being Conciliatory: Remarks about Mellon's Model for Hospital Chaplaincy Work in Multi-Faith Settings" in: *Christian Bioethics* (2003, vol.9, no.1); pg. 71

3) Pope John Paul II; "The Gospel of Life" in: *Last Rights? Assisted Suicide and Euthanasia Debated* (1998), Eerdmans; pg.223–

4) Pope John Paul II describes this as a 'new cultural climate' characterized by attacks against life by a corrupt view of individual rights and the enactment of laws that deviate from basic constitutional law supported by the assistance of health care systems. ((See Endnote 4, pg.226.))

5) Singer, Peter, 1998, "Rethinking Life and Death: A New Ethical Approach" ((See Endnote 4, pg.171–))

6) Kevorkian, Jack, 1998, "A Fail-Safe Model for Justifiable Medically Assisted Suicide", ((See Endnote 4, pg.263–))

7) Editorial; "Death by Default" in *Christianity Today* (5 February 2001), pg. 26–

8) Drane, James F.; *More Humane Medicine: A Liberal Catholic Bioethics* (2003) Attenborough

9) Clowney, Edmund; "Jesus Christ and the Lostness of Man" in: *Jesus Christ: Lord of the Universe, Hope of the World* (1974), InterVarsity Press; pg. 53–

10) Barth, Karl; *The Epistle to the Romans* (1933; reprint 1976, 6th edition), Oxford

11) Vaux, Kenneth L., & Vaux, Sara A.; *Dying Well* (1996) Abingdon

12) Callahan, Daniel; *Is Research a Moral Obligation?* (2003) http://www.bioethics.gov/background/callahan_paper.html

13) Haseltine, William, quoted by Fisher, Lawrence M.; "The Race to Cash in on the Genetic Code" in: *The New York Times* (29 August 1999), section 3, pg. 1

SUPERLONGEVITY WITHOUT OVERPOPULATION

Max More, Ph.D.

Proponents of superlongevity (indefinitely extended life spans) have been making their case for the possibility and desirability of this change in the human condition for decades. For just as long, those hearing the arguments for superlongevity have deployed two or three unchanging, unrelenting responses. The question: "But what would we do with all that time?" is one of them. Another is the "But death is natural!" gambit. The final predictable response is to conjure up the specter of overpopulation. Despite strong downward trends in population growth since this issue gained visibility in the 1960's, the third concern remains an impediment.

Paul Ehrlich's 1968 bestseller, *The Population Bomb* [1], ignited a trend in which alarmists routinely ignored data and reasonable projections to scare the public. Those of us who see achieving the indefinite extension of the human life span as a central goal naturally find this behavior quite irritating. If baseless fear wins out, we will gain little from our personal programs of exercise, nutrition, or supplementation. Widespread fear leads to restrictive legislation – legislation that in this case could be deadly. Although the volume has been turned down a little on the population issue, it continues to reverberate and deserves a response. The purpose of this essay is to address the essential concerns, provide current facts, and dispel the errors behind the overpopulation worries.

VALUES FIRST

As I will show, we have little reason to fear population growth with or without extended lives. However, to bring into focus an ethical issue, I will pretend for a moment that population growth is or will become a serious problem. Would this give us a strong reason for turning against the extension of human lifespan?

No. Opposing extended life because, eventually, it might add to existing problems would be an ethically irresponsible response. Suppose you are a doctor faced with a child suffering from pneumonia. Would you refuse to cure the child because she would then be well enough to run around and step on the toes of others? On the contrary, our responsibility lies in striving to live long and vitally while helping others do the same. Once we are at work on this primary goal, we can focus more energy on solving other challenges. Long, vital living at the individual level certainly benefits from a healthy physical and social environment. The superlongevity advocate would want to help find solutions to any population issues. But dying is not a responsible or healthy way to solve anything.

Besides, if we take seriously the idea of limiting life span so as to control population, why not be more proactive about it? Why not drastically reduce access to currently commonplace medical treatments? Why not execute anyone reaching the age of seventy? Once the collective goal of population growth is accepted as overriding individual choices, it would seem hard to resist this logic.

It is How Many, Not How Long, That Matters?

Limiting population growth by opposing life extension not only fails the ethical test, it also fails the pragmatic test. Keeping the death rate up simply is not an effective way of slowing population growth. Population growth depends far more on how many children families have, as opposed to how long people live. In mathematical terms, longer life has no effect on the exponential growth rate. It only affects a constant of the equation. This means that it matters little how long we live after we have reproduced. Compare two societies: In country A, people live on average only to 40 years of age, each family producing 5 children. In country B, the life span is 90 years but couples have 4 children. Despite the much longer life span in country B, their population growth rate will be much lower than that of country A. It makes little difference over the long term how many years people live after they have had children. The population growth rate is determined by how many children we have, not how long we live.

Even the short-term upward effect on population due to a falling death rate may be cancelled by a delay in child bearing. Many women in developed countries choose to bear children by their early 30's because the obstacles to successful pregnancy grow as they age. As the last few decades have already shown, extending the fertile period of women's lives would allow them to put off having children until later, until they have developed their careers. Not only will couples have children later, we can expect them to be better positioned financially and psychologically to care for them.

Almost certainly, the first truly effective technologies to extend the maximum human life span will come with a significant cost of human development and application. As a consequence population effects would first be felt in the developed countries. This points to another flaw in the suggestion that extended longevity will dramatically boost population growth. The fact is, superlongevity in the developed nations would have practically no global or local population impact. The lack of global impact is a consequence of the small and falling share of the global population accounted for by the developed nations. No local population boom drama can realistically be expected because these countries are experiencing very low, zero, or negative population growth:

The share of the global population accounted for by the developed nations has fallen from 32 percent in 1950 to 20 percent currently and is projected to fall to 13 percent in 2050. [2] If we look just at Europe, we see an even more remarkable shrinkage: In 1950, Europe accounted for 22 percent of the global population. Currently it has fallen to 13 percent, and is projected to fall to 7 percent by 2050. [3] To put this in perspective, consider that the population of Africa at 749 million is now greater than that of Europe at 729 million, according to UN figures. Europe's population growth rate of just 0.03 per cent will ensure that it will rapidly shrink relative to Africa and other developing areas.

In Eastern Europe, population is now shrinking at a rate of 0.2 percent. Between now and 2050, the population of the more developed regions is expected to change little. Projections show that by mid-century, the populations of 39 countries will be smaller than today. Some examples: Japan and Germany 14 percent smaller; Italy and Hungary 25 percent smaller; and the Russian Federation, Georgia and Ukraine between 28–40 percent smaller. [3]

For the United States (whose population grows faster than Europe), the bottom line was summed in a presentation to the President's Council on Bioethics by S.J. Olshansky who "did some basic calculations to demonstrate what would happen if we achieved immortality today." The bottom line is that if we achieved immortality today, the growth rate of the population would be less than what we observed during the post World War II baby boom. [4]

Low fertility means that population trends in the developed regions of the world would look even milder if not for immigration. As the 2000 Revision to the UN Population Division's projections says: "The more developed regions are expected to continue being net receivers of international migrants, with an average gain of about 2 million per year over the next 50 years. Without migration, the population of more developed regions as a whole would start declining in 2003 rather than in 2025, and by 2050 it would be 126 million less than the 1.18 billion projected under the assumption of continued migration."

All things considered, countries fortunate enough to develop and make available radical solutions to aging and death need not worry about becoming overpopulated. In an ideal scenario, life extension treatments would rapidly plunge in cost, making them affordable well beyond the richest nations. We should therefore look beyond the developed nations and examine global population trends in case a significantly different picture emerges.

Global Population Flatlining

We have seen that we have no reason to hesitate in prolonging life even if population were to grow faster due to higher fertility rates. But does the developing world, with or without cheap, ubiquitous life extension, have much to fear from a population explosion? Are populations growing out of control in those regions? The fad for popular books foretelling doom started in the 1960's, at the tail end of the most rapid increase in population in human history. Since then, the poorer countries, well below us in the development cycle, have also been experiencing a drastic reduction of population growth. This is true despite major relative life extension – the extra decades of life bestowed by medical intervention and nutrition.

Taking a global perspective, the numbers reveal that the average annual population growth rate peaked in 1965–1970 at 2.07 percent. Ever since then, the rate of increase has been declining, coming down to 1.2 per cent annually. That means the addition of 77 million people per year, based on an estimated world population of 6.1 billion in mid-2000. [3]

A mere six countries account for fully half of this growth: India for 21 percent; China for 12 percent; Pakistan for 5 percent; Nigeria for 4 percent; Bangladesh for 4 percent, and Indonesia for 3 percent. China has markedly reduced the average number of births per woman over the last 50 years from six to 1.8. Starting from the same birth rate at that time, India has fallen much less, although still almost halving the rate to 3.23 percent. If these trends continue up to 2050, India's population will exceed that of China. [5]

Despite the fecundity of these top people-producers, the overall picture is an encouraging one:

> The total fertility rate for the world as a whole dropped by nearly two-fifths between 1950/55 and 1990/95 – from about 5 children per woman down to about 3.1 children per woman. Average fertility in the more developed regions fell from 2.8 to 1.7 children per woman, well below biological replacement. Meanwhile total fertility rates in less developed nations fell by 40 percent, falling from 6.2 to 3.5 children per woman. [6]

We can expect population growth to continue slowing until it reaches a stable size. What size will that be? No one knows for sure, but the best UN numbers indicate that population may peak at as low as 8 billion people, with a medium projection of 9.3 billion and an upper limit projection of

10.9 billion. [2;7] The medium projection also points to global population peaking around 2040 and then starting to fall.

I wrote the first version of this paper in 1996. In revising it, I found it interesting that, less than a decade ago, the higher projection allowed for 12 billion or more. Demographers had continued their long tradition of over-estimating population growth. This effect seems to have been reduced, but take all projections (especially those longer than a generation) with a healthy dose of skepticism.

FORCES OF POPULATION DECELERATION

Why, though, should we expect people in less developed countries, even given contraceptives, to continue choosing to have smaller families? This expectation is not merely speculation based on recent trends. Sound economic reasoning explains the continuing trend, and makes sense of why the poorer nations are only just beginning to make the transition to fewer births.

Decelerating population growth appears to be an inevitable result of growing wealth. Early on in a country's developmental curve, children can be regarded as 'producer goods' (as economists would say). Parents put their children to work on the farm to generate food and revenue. Very little effort is put into caring for the child: no expensive health plans, special classes, trips to Disneyland, X-Men action figures, or mounting phone bills. As we become wealthier, children become 'consumer goods'. That is, we look on them more and more as little people to be enjoyed and pampered and educated, not beasts of burden to help keep the family alive. We spend thousands of dollars on children to keep them healthy, entertain them, and educate them. We come to prefer fewer children to a vast mob of little ones. This preference seems to be reinforced by changing tastes resulting from improved education. The revenue vs. expense equation for extra children further shifts toward having fewer offspring as populations become urbanized. Children cost more to raise in cities and can produce less income than in the country.

Fertility declines for another reason: As poorer countries become wealthier, child mortality falls as a result of improved nutrition, sanitation, and health care. Reduced child mortality in modern times can come about even without a rise in income. People in poorer countries are not stupid; they adjust their childbearing plans to reflect changing conditions. When child death rates are high, research has shown that families have more children to ensure achieving a given family size. They have more children to make up for deaths, and often have additional children in anticipation of later deaths. Families reduce fertility as they realize that fewer births are needed to reach a desired family size. Given the incentives to have fewer children as wealth grows and urbanization proceeds, reduced mortality leads to families choosing to reduce family size.

Economic policy helps shape childbearing incentives. Many of the same people who have decried population growth have supported policies guaranteed to boost childbirths. More than that, they boost childbearing among those least able to raise and educate children well. If we want to encourage people to have more children, we should make it cheaper for them to do so. If we want to discourage fertility, or at least refrain from pushing it up, we should stop subsidizing it. Subsidies include free education (free to the parents, not to the tax-payers), free child health care, and additional welfare payments to women for each child they bear. If parents must personally bear the costs of having children, rather than everyone else paying, people will tend to have just the number of children for whom they can assume financial responsibility.

Even if there were a population problem in a few countries, extending the human life span would worsen the problem no more than would improving automobile safety or worker safety, or reducing violent crime. Who would want to keep these deadly threats high in order to combat population growth? If we want to slow population growth, we should focus on reducing births, not on raising or maintaining deaths. If we want to reduce births, we might voluntarily fund programs to provide contraceptives and family planning to couples in poorer countries. This will aid the natural developmental process of choosing to have fewer children. Couples will be able to have children by choice, not by accident. Women should also be encouraged to join the modern world by gaining the ability to pursue vocations other than child-raising.

'Overpopulation' Distracts from Real Problems

Major downward revisions in population growth – throughout the UN's sixteen rounds of global demographic estimates and projections since 1950 – have drained the plausibility of any overpopulation-based argument against life extension. We can better understand the real problems that are talked about in relation to overpopulation instead as issues of poverty. Poverty, in turn, results not from having too many people, but from several major factors including political misrule, continual warfare, and insecurity of property rights.

As Bjorn Lomborg points out, we find many of the most densely populated countries in Europe. The region with the highest population density, Southeast Asia, has about same number of people per square mile as the United Kingdom. Although India has a large, growing population, it also has a population density far lower than that of The Netherlands,

Belgium, or Japan. Lomborg also notes that Ohio and Denmark are more densely populated than Indonesia. [3]

We should also recognize that most population growth takes place in urban areas, which provide a better standard of living. As a result, most of this planet's landmass will not be more densely populated than it is today. Over the next three decades, we can expect to see almost no change in the rural population of the world and, by 2025, 97% of Europe will be less densely populated than today. [8] We should celebrate the urbanization trend since even the urban poor thrive better than they would in the country. The causes of this include better water supplies, sewage systems, health services, education, and nutrition. [9] Oddly enough, serious infectious diseases like malaria are less threatening the closer buildings are together (and so the smaller the space for swampy areas beloved of mosquitoes and flies). [10]

SUSTAINABILITY AND THE GREAT RESTORATION

The future could be far brighter than the eco-doomsters have long portrayed it. As Ronald Bailey [11] reports:

> Jesse Ausubel, director of the Program for the Human Environment at Rockefeller University, believes the 21st century will see the beginning of a 'Great Restoration' as humanity's productive activities increasingly withdraw from the natural world.

If world farmers come to match the typical yield of today's US corn growers, ten billion people could eat amply while requiring only half of today's cropland. This is one way in which technological advance in farming will allow vast expanses of land to revert to nature. Transgenic crops could also multiply today's production levels while solving several significant environmental challenges. [12]

Visions that emphasize human ingenuity and opportunity have a far more impressive historical record than those that emphasize human passivity and helplessness. Paul Ehrlich is a classic case of the latter type and you have only to browse his dark, alarming books to recognize how consistently bad he has been at making environmental predictions. In a 1969 article, Ehrlich predicted the oceans dead from DDT poisoning by 1979 and devoid of fish; 200,000 deaths from 'smog disasters' in New York and Los Angeles in 1973; U.S. life expectancy dropping to 42 years by 1980 because of pesticide-induced cancers, and U.S. population declining to 22.6 million by 1999. [13] Ehrlich famously lost a ten year bet against cornucopian economist Julian Simon (and refused to renew the bet). [14] In 1974, Ehrlich recommended stockpiling cans of tuna due to the certainty of protein shortages in the USA. And so on.

As Bailey explains [13], contrary to Ehrlich:

> Instead, according to the United Nations, agricultural production in the developing world has increased by 52 percent per person since 1961. The daily food intake in poor countries has increased from 1,932 calories, barely enough for survival, in 1961 to 2,650 calories in 1998, and is expected to rise to 3,020 by 2030. Likewise, the proportion of people in developing countries who are starving has dropped from 45 percent in 1949 to 18 percent today, and is expected to decline even further to 12 percent in 2010 and just 6 percent in 2030. Food, in other words, is becoming not scarcer but ever more abundant. This is reflected in its price. Since 1800 food prices have decreased by more than 90 percent, and in 2000, according to the World Bank, prices were "lower than ever before".

A reading of economic and social history quickly makes one thing plain: throughout history people have envisaged overpopulation. Even the great nineteenth century social scientist W. Stanley Jevons in 1865 claimed that England's industrial expansion would soon cease due to the exhaustion of the country's coal supply. [15] However, as shortages developed, prices rose. The profit motive stimulated entrepreneurs to find new sources, to develop better technology for finding and extracting coal, and to transport it to where it was needed. The crisis never happened. Today, the USA has proven reserves sufficient to last hundreds or thousands of years. [16] If one resource does begin to run low, rising prices will encourage a switch to alternatives. Even a vastly bloated population cannot hope to exhaust energy supplies. (Solar energy and power from nuclear fission and soon fusion are practically endless.) So long as we have plentiful energy we can produce substitute resources and even generate more of existing resources, including food. Even if population were to grow far outside today's highest projections, we can expect human intelligence and technology to comfortably handle the numbers.

Human intelligence, new technology, and a market economy will allow this planet to support many times the current population of 6.2 billion – it can support many more humans than we are likely to see, given trends toward lower birth rates. Many countries, including the USA, have a rather low population density. If the USA's population were as dense as Japan – hardly a crowded place overall – our population would be 3.5 billion rather than 265 million. If the USA had a population density equal to that of Singapore, we would find almost 35 billion people here, or almost seven times the current world population. New technologies, from simple improvements in irrigation and management to current breakthroughs in genetic engineering should continue to improve world food output. Fewer people are starving despite higher populations. This does not mean they are feeling satisfied. Millions still go hungry or are vulnerable to disruptions in supply. We need to push to remove trade barriers, abolish price controls on agriculture (which discourage production and investment), and pressure governments engaging in warfare and collectivization to change their ways.

POLLUTION

Nor should we expect pollution to worsen as population grows. Contrary to popular belief, overall pollution in the more developed countries has been decreasing for decades. In the USA, levels of lead have dropped dramatically. Since the 1960's levels of sulfur dioxide, carbon monoxide, ozone, and organic compounds have fallen despite a growing population. Air quality in major urban areas continues to improve, and the Great Lakes are returning toward earlier levels of purity. [17] This is no accident. As we become wealthier, we have more money to spare for a cleaner environment. When you are longing for food, shelter, and other basics, you will not spare much thought for the environment. So long as mechanisms exist for converting desires for cleaner air and water and space for recreation into the things themselves, we can expect it to happen.

Most effective at spurring the positive changes are markets – price signals creating incentives for moves in the right direction. If polluters must pay for what they produce because their activity intrudes on the property rights of others, they will search for ways to make things with less pollution. Pollution problems do exist. Most of them can be traced to a failure to enforce private property rights, so that resources are treated as free goods that need not be well managed. Fishing in unowned bodies of water is an example of this. The desertification of collectively or government owned land in Africa is another. We can be reasonably confident that the trend towards less pollution with greater population will continue. However, complacency is out of place. We should press for responsible management of resources by privatizing collectively owned resources to create incentives for sound management and renewal.

So long as we continue to allow freedom to generate more wealth and better technology, we can expect pollution to continue abating. More efficient recycling, production processes that generate fewer pollutants, and better monitoring and detection of polluters, along with economic incentives making each producer responsible for their output, will allow us to continue improving our environment even as population grows. Assuming that we achieve complete control of matter at the molecular level, as expected by nanotechnologists, we will have the keys to production without pollution. Another product of molecular manufacturing will be the disappearance of most large-scale, clumsy machinery. Less and less land will need to be used for manufacturing equipment, making more room for people to enjoy. Some manufacturing will be moved into space. The result of these and other changes (some of which are already underway) will be the freeing of the Earth from unwanted, but previously necessary, means and by-products of manufacturing.

The population issue raises numerous factual, economic, and ethical concerns. I urge the interested reader to check into the sources listed in the References, especially the essays by Jesse Ausubel [18] and the books by Bailey, Lomborg, and Simon. [3;19;20–25] I have only sketched lines of thinking showing that we would be severely misguided not to push for extended life out of fear of overpopulation. Let us move full speed ahead with extending life span: Once we have vanquished aging, I would expect other threats to life, such as war and violent crime, will become even less acceptable. We can look forward to a long-lived society better off than previous generations; not only in economic well being, but also in security of life and health.

References

1) Ehrlich, Paul R; *The Population Bomb* (1968); Sierra Club-Ballantine

2) World Population Prospects: *The 2000 Revision* (2001a); United Nations Publications

3) Lomborg, Bjorn; *The Skeptical Environmentalist: Measuring the Real State of the World* (2001); Cambridge University Press

4) Olshansky, SJ; "Duration of Life: Is There a Biological Warranty Period?" in: *The President's Council on Bioethics* (2002) Washington, DC http://www.bioethics.gov/transcripts/dec02/session2.html

5) World Population Prospects: *The 2000 Revision, Additional Data* (2001c); United Nations Publications

6) Eberstadt, Nicholas; "Population, Food, and Income: Global Trends in the Twentieth Century" in: *Bailey* (1995)

7) World Population Prospects: *The 2000 Revision, Annex Tables* (2001b); United Nations Publications

8) World Urbanization Prospects: *The 1996 Revision* (1998); United Nations Publications

9) The Progress of Nations (1997) UNICEF http://www.unicef.org/pon97/

10) Miller, Jr. Tyler G; *Living in the Environment: Principles, Connections, and Solutions* (1998); Wadsworth Publishing Company

11) Bailey, Ronald; "The End Is Nigh, Again" in: *Reason* (2002); June 26

12) Rauch, Jonathan; "Will Frankenfood Save the Planet?" in: *The Atlantic Monthly* (2003); October

13) Bailey, Ronald; *Eco-Scam* (1993); St. Martin's Press

14) http://www.overpopulation.com/faq/People/julian_simon.html

15) Jevons S; *The Coal Question: An inquiry concerning the progress of the nation and the probable exhaustion of our coal mines* (1865); Kelley Publishers

16) http://www.eia.doe.gov/

17) Taylor, B et al. "Water Quality and the Great Lakes" in: *Michigan's Opportunities and Challenges: Msu Faculty Perspectives*, Michigan in Brief: 2002–03. Public Sector Consultants, Inc.

18) Ausubel, Jesse; "The Great Restoration of Nature: Why and How" in: *Challenges of a Changing Earth* (2002); pg.175–182

// *Proceedings of the Global Change Open Science Conference*, Amsterdam, Netherlands (2001, 10–13 July) edited by Steffen, W & Jaeger, J & Carson, DJ & Bradshaw C; Springer http://phe.rockefeller.edu/sthubert/hubert.pdf

// Ausubel, Jesse; "Where is Energy Going?" in: *The Industrial Physicist* (2000); http://phe.rockefeller.edu/IndustrialPhysicistWhere/where.pdf

19) *The True State of the Planet* (1995); edited by Bailey, Ronald; The Free Press

20) Simon, Julian L; "Resources, Population, Environment: An Over-Supply of False Bad News" in: *Science* (1980, Vol. 280); pg.1431–1437

21) Simon, Julian L; *The Ultimate Resource* (1981); Princeton University Press

22) Simon, Julian L; "Forecasting the Long-Term Trend of Raw Material Availability," in: *International Journal of Forecasting* (1985, Vol. 1); pg.85–109

23) Simon, Julian L; *Population Matters* (1990); N.J.: Transaction

24) Simon, Julian L; "Bunkrapt: The Abstractions that lead to scares about resources and population growth," in: *Extropy* (1993, Vol. 11); Summer/Fall 1993, pg.34–41

25) *The Resourceful Earth* (1984); edited by Simon, Julian L & Kahn, Herman; Basil Blackwell, Inc.

EMANCIPATION FROM DEATH

Mike Treder

In the time that it takes you to read this sentence, at least 10 real people will die, some of them helpless children, and some in horrible pain. Every single day 24,000 people die of starvation; 6,000 children are killed by diarrhea; 2,700 children are killed by measles; and 1,400 women die in childbirth. [1]

All told, more than 150,000 humans will lose their lives today. Some of them will be elderly, of course, but why should that be a death sentence? Even worse, tens of thousands of youthful adults and children will be lost tomorrow – and the next day, and the day after that – to preventable or curable illnesses simply because treatment is not available to them. Must we accept this daily horror? Is it really necessary? I believe it is time we start fighting back; the good news is we are.

Each day significant progress is being made to defeat disease and reduce suffering. In addition, work is well underway to understand the aging process and someday eliminate it. As Robert Ettinger has said: "Being born is not a crime, so why must it carry a sentence of death?" [2] In the appeals court of science and technology, the summary execution of every human being may soon be overturned, hopefully within your lifetime.

UPSETTING THE 'NATURAL' ORDER

"This is hubris," some tell us. "Death is natural, and we must not play God." [3] Yet ever since the earliest human donned an animal skin, we have used our native resourcefulness and creativity to enhance our security, comfort, and efficacy; from the loincloth to the toga to the modern suit, from Ben Franklin's bifocals to contact lenses to laser eye surgery.

In modern marketing, products are commonly promoted as 'natural'. But what is natural? And what is unnatural? By the most precise definition, everything that occurs in our world – whether synthetic or not – is natural, because humans are a part of nature and therefore the products of our hands – or our machines – are also part of nature. That is not, however, the meaning of 'natural' that most people intend. Rather, they are referring to products, events, or occurrences not made or caused by human beings. Thus, milk would be classified as 'natural', while Kool-Aid would not. (Never mind that the milk we buy in cartons at the store has been pasteurized, homogenized, and vitamin fortified.) Less trivial debates surrounding the word 'natural' arise when considering enhancements that might be made to human beings, especially when we talk of defeating death. It is interesting to note that numerous other scientific measures to improve the human condition have initially been scorned as unnatural and intolerable by many, only to be later accepted almost universally. Examples include anesthesia, blood transfusions, vaccinations, birth control pills, and organ transplants. Consider what our world might be like without these and hundreds of other improvements that may not fit the popular definition of 'natural'.

Tooth decay is natural – should dentistry be outlawed? Polio is natural – should we ban the Sabin vaccine? Cholera is natural – should we allow epidemics to rage unchallenged? Death is natural – must it continue to wreak its dreadful havoc? Clearly this is foolishness. Of course we should use all available means to better human life. We have been doing it for ages with fire, farming, steam, electricity, antibiotics, vaccines, dental prosthesis, organ transplants, etc.; and we should not stop now. If modern science and technology can safely improve the human condition by overcoming natural limits, including aging and death, then they should be used to this end. Determining whether something is good or bad simply by asking whether or not it is natural does not follow common sense.

All this is not to say that we should ignore the moral and ethical challenges that confront us. Questions of safety, propriety, individual choice, and societal responsibility must not be dismissed, but must be considered gravely and at length. Issues of overpopulation, reproductive rights, resource distribution, and environmental impact must be addressed forthrightly. This can only be done, however, in an atmosphere of openness and progressive thinking.

EMANCIPATION FROM DEATH

For those who still believe that opposing death is somehow wrong or unnatural, please remember that opposition to human slavery was also once considered crazy and dangerous. Arthur C. Clarke has written:

> Every revolutionary idea evokes three stages of reactions: At first people say, "It's completely impossible." Then they say, "Maybe it's possible to do it, but it would cost too much." Finally they say, "I always thought it was a good idea." [4]

Clarke's amusing observation is exactly on target. Staying with the analogy of human slavery, note that throughout most of history (and, no doubt, prehistory), it has been common for some humans to own other humans. [5] The movement toward recognition of freedom as a fundamental human right is relatively recent. During the drafting of the U.S. Constitution, its framers debated how to handle the so-called 'slave issue'. This was at a time, recall, when many nations, particularly those in Western Europe, had already abolished the practice. Although a large number of American leaders abhorred slavery, totally eliminating it was widely considered to be "completely impossible". [6] As U.S. history proceeded and opposition to slavery grew, the debate changed to one of practicality. People said, "Maybe it's possible to do it, but it would cost too much." A few generations after a bloody, costly, destructive and painful civil war, descendants of Americans who once owned other humans would say about abolition, "I always thought it was a good idea." When the time comes – and it will – that humans are no longer enslaved by death, leadership on this issue will be recognized for what it is: courageous, honest, and humane.

Biotechnology and nanomedicine may hold the promise for us to live forever free from illness, disease, and physical disability; always youthful and vigorous; free to do whatever we want with our lives; liberated from the constraints of ill health and physical frailty.

In addition to the obvious hope of living without death in human bodies, there are numerous other ways we can imagine extending our lives. One way is to inject our personality into a virtually indestructible robot. This might be done by physically relocating the brain from our frail, vulnerable body and implanting it into a robot; but more likely it would be done by making a digital copy of our brain and downloading all the information into the robot. This method has the advantage of being able to preserve a backup copy of our personality, as insurance against the remote possibility that something catastrophic might destroy our robot body. This really would make us effectively immortal, as we could store copies of ourselves in places all over the solar system, the galaxy, or eventually even beyond.

SIMULATING IMMORTALITY

It is a loathsome and cruel trick that nature takes such an exquisitely wondrous creation as the human brain and imprisons it inside the weak, inefficient, fragile, and short-lived structure that is the human body. Our bodies may be beautiful, but they are unacceptably ephemeral.

The body you now inhabit, however remarkable it may be, is not the product of intelligent design. It was not created for any purpose other than survival and reproduction. We are conditioned by social and biological forces to favor the appearance of the human form and to be attracted by its outlines and contours. It is therefore a natural reaction – although not necessarily a rational one – for us to be repelled by any substantial deviation from the standard model. That's why most of us cringe (at least inwardly) at the sight of a person with a disfigured face or missing limbs. It also explains why many people are repelled by the thought of replacing the natural human body with one of artificial design and creation.

And yet, why not? The body we were given by nature is the result of millions of years of meandering and directionless change. It is the product of a tortuous, cumbersome, slow and dumb process called evolution. The human body was not designed for our optimum enjoyment and benefit; it became as it is now basically by accident. Nature, given its leeway, would continue to blindly experiment with us. Following the random cues of genetic mutation, our bodies would slowly evolve, gradually becoming something different.

In contrast, we humans are highly intelligent creatures and have reached the point where we can take the future development of our bodies into our own hands. Using our minds and the marvelous tools we are now making, we can produce a new form – or many new forms – for the body. We can design to suit our own purposes and preferences.

In the past, engineers developing new prototypes for aircraft, automobiles, or ocean liners would create scale models and then evaluate the performance of their concepts in wind tunnels or other testing media. Modern engineers find it easier, cheaper, and more effective to do the same type of testing in simulated environments. Using powerful computers and highly sophisticated software programs, they can learn precisely how their creations will perform under a variety of conditions.

As a way of experimenting with possible new designs for your posthuman body, you will likely do the same thing. Instead of going to the trouble of building your new body molecule by molecule and then determining whether it is satisfactory, you can create a simulation in a virtual reality environment and test it there. The exciting difference is that you will not be limited to observing the simulation as are today's engineers. Rather, you will be able to inhabit your virtual body and know firsthand how it will react, perform, and feel.

The next step is obvious. If the simulation is powerful enough, the experience of occupying the simulated body should be indistinguishable from conventional physical reality; it will be virtually the same – hence the name 'virtual reality'. Then why not just live there? Assuming you can have all the experiences of the 'real' world – plus many more that you could never have – and that you will still be able to see and touch and interact with the people you love, why not just stay?

Many humans today might recoil from the idea of living only within a virtual realm. But from a philosophical perspective, there is truly no difference between the experience of inhabiting a sufficiently advanced simulation, and the everyday life that we experience today. Consider this: our current physical bodies can be thought of as organic robots. They go out into the physical world, carrying a brain/mind/personality/identity around inside. My organic robot body sees, hears, touches, smells, and tastes for me; it transmits those experiences to my brain through electrical pathways; parallel processing computation within my neurons and synapses results in a pattern of thought so complex and elegant that it generates meta-cognition, or self-awareness. I think it is 'me' that is out there in the world enjoying direct sensory experiences, but it is not!

The part of me that is really me – the part that is my consciousness and my personality – can never have such direct experiences. Gray matter has neither hands, nor eyes, nor ears, nor mouth, nor nose. My brain must rely on an indirect interface to apprehend 'reality.' That interface can be the physical body I now inhabit, it could be a tele-robot exploring the surface of Mars, or it could be a substrate of computation providing a 'simulated' environmental experience.

The point is that everything we experience is simulated. Nothing is immediate. Over the next few decades, as we spend more and more time in virtual environments, our definition of reality will change. It is conceivable that within a century or less, many human personalities may be living full-time in cyberspace, inhabiting myriad simulations. They will undoubtedly discover new sensations and emotions we cannot even comprehend. Will their lives be less 'real' than ours today?

It seems likely that millions of people, if not billions, will make just that choice. Does this sound like science fiction?

Perhaps so, but current trends in computing technology suggest this could start to become a reality within as little as 20 or 30 years from now. [7]

THE UNIMAGINABLE FUTURE

Here's another fascinating question. If living in one body is good, why not have two? If two is good, why not have three or four or five? Why not five hundred or five million?

World Future Society President Edward Cornish has said,

> In our most imaginative fantasies, we cannot anticipate all the extraordinary possibilities of the future for us humans and whatever creatures come after us. The wildest speculations of today may be the facts of tomorrow, and our human potential is not only greater than we think but greater than we can think. [8]

Imagine for a moment inhabiting multiple bodies; not merely having a variety of bodies to choose from, like suits of clothes in a closet, but being in many different bodies at the same time. One of the bodies might be the one you were born with; others might be duplicates or clones; some could be substantially different, perhaps designed to fit a specific environment; a majority of them will probably be robot bodies or virtual bodies.

What will be your experience of personal identity when your consciousness is spread over many different substrates? Will you still be you? Will you choose to maintain, as much as possible, simultaneous awareness in all the bodies at once? Or will it be preferable to allow your bodies to function autonomously with an occasional, perhaps daily, synchronization of your experiences and realignment of your identity?

It is even conceivable that in the future we will be able to simulate the personalities of people from the past – whether celebrities, historical figures, or loved ones – and relate directly with them. It is also possible that you might (with their permission, of course) choose to integrate one or more of these identities into your own.

You may also someday accept the invitation to become part of a meta-being by subsuming your identity (or maybe a copy of your identity) into theirs. Some have speculated that the long-term evolution of posthumans must follow this pathway into integrated immortal super-beings. [9]

Whatever happens, it is clear that the future will be much stranger – and far more wonderful – than we have ever imagined.

References

1) United Nations Demographic Yearbook, 2000 // http://www.disasterrelief.org/Disasters/001023hungerreport/

2) http://www.wordiq.com/definition/Immortality

3) McKibben, Bill; *Enough: Staying Human in an Engineered Age* (2003); Times Books ; // Kass, Leon; *Beyond Therapy: Biotechnology and the Pursuit of Happiness* (2003); Regan Books

4) Bova, Ben; *Immortality: How Science Is Extending Your Life Span and Changing the World* (1998); William Morrow & Company; pg.183

5) Meltzer; Milton; *Slavery: A World History* (1993); Da Capo Press

6) Hummel, JR; *Emancipating Slaves, Enslaving Free Men: A History of the American Civil War* (1996); Open Court Publishing Company

7) Kurzweil, Ray; *The Age of Spiritual Machines: When Computers Exceed Human Intelligence* (1999); Viking / Penguin Books

8) Cornish, Edward; *Futuring: The Exploration of the Future* (2004); World Future Society; pg.121

9) Tipler, Frank J; *The Physics of Immortality: Modern Cosmology, God and the Resurrection of the Dead* (1994); The New York Times Company

The Self-Defeating Fantasy

Eric S. Rabkin, Ph.D.

In our oldest tale, *The Epic of Gilgamesh* from the 3rd millennium B.C.E., the hero learns

> a secret thing [a mystery of the gods]. There is a plant that grows under the water, it has a prickle like a thorn, like a rose; it will wound your hands, but if you succeed in taking it, then your hands will hold that which restores his lost youth to a man (pg. 116). [1]

To retrieve immortality, Gilgamesh weights himself with stones and plunges into the life-offering, death-threatening water. But

> deep in the pool there was lying a serpent, and the serpent sensed the sweetness of the flower. It rose out of the water and snatched it away, and immediately it sloughed its skin and returned to the well. Then Gilgamesh sat down and wept, the tears ran down his face. I found a sign and now I have lost it (pg. 117). [1]

Italo Calvino has written that

> the ultimate meaning to which all stories refer has two faces: the continuity of life, the inevitability of death (pg. 259). [2]

We see both in this founding tragedy, for nature in the form of the snake returns to the pool, able to escape its corporeality and renew it, while humanity in the form of Gilgamesh can only return to the dusty city of Uruk, well built it is true, but ultimately a feeble defense against death. Nonetheless, many still hope for immortality, feeling, like Dostoevsky, that

> if you were to destroy in mankind the belief in immortality, not only love but every living force maintaining the life of the world would at once be dried up (581:19). [3]

Yet our fictions often tell us that immortality is best only as a hope and never as an actuality, for, despite its venerable, obvious, and intimate appeal, the fantasy of immortality masks a terrible reality.

The clearest warnings against immortality, some might suggest, are really warnings against hubris, foolishness, and disobedience. The Cumaean Sybil, adored by Apollo, is granted a thousand years of life, but because she spurns the love of the god, he withholds eternal youth and she suffers on and on. Tithonus, beloved of Eos, the Goddess of Dawn, is granted immortality but forgets to ask for eternal youth, so he ages forever in what Tennyson has him call "cruel immortality". Prometheus is by nature an immortal, but for having stolen fire for humanity, his immortality becomes an eternity of suffering. One could say that immortality in these cases is no worse in itself than gold is in the story of Midas: a fine thing in its proper place, but ironic, indeed tragic, when corrupted. The apotheoses of Greek heroes and Hebrew prophets would seem to corroborate this positive view of immortality, as would the irony of so fine a state leading not to happiness but to horror. However, can we find an immortality that does not suffer such fatal defects?

It is often said that the central promise of Christianity is immortality: "I am the resurrection, and the life: he that believeth in me, though he were dead, yet shall he live: And whosoever liveth and believeth in me shall never die" (Jn 11:25–26). History shows that this promise has much appeal, but, curiously, we have very few glimpses of what it would mean to live this perfect immortality. In *Man and Superman*, George Bernard Shaw clearly prefers hell, "the home of the unreal and of the seekers for happiness" to heaven "the home of the masters of reality, and [earth] [...] the home of the slaves of reality" (pg. 139). [4]

This matter of masters and slaves brings us back to the issue of disobedience. Milton wrote in the opening lines of *Paradise Lost*:

> Of Man's First Disobedience, and the Fruit / Of that Forbidden Tree, whose mortal taste / Brought Death into the World, and all our woe. [5]

If Jesus is the new Adam, then his redemption of us is a return to Edenic obedience, for, as Milton clearly says, death and disobedience stand against life and, one presumes, obedience. Yet a heaven of perfect obedience, when concretely realized, hardly seems human happiness, so dependent is our happiness on notions of individual freedom and of desire. Adam, like Gilgamesh, lost immortality through the intervention of a serpent. One supposes that in heaven there are no serpents, nor any dangers, nor even the sexuality that such serpents in part represent. Shaw's heaven, like St. John's, suffers from what Arthur C. Clarke calls "the supreme enemy of all Utopias –boredom" (pg. 75). [6]

The paradigmatic benevolence of Christianity, the compensation, as it were, for Original Sin and the Flood, is God the Father projecting himself into the mortal reality of Jesus. For believing Christians, of course, this is a unique and pivotal event in human history; I do not mean to comment on such beliefs. But in fiction, the willingness to accept mortality is by no means rare, and, where there is no promise of life-everlasting, as there is not, say, for Sidney Carton when he takes Charles Darnay's place at the guillotine at the end of Charles Dickens' *A Tale of Two Cities*. [7] Such mortality is the measure of human, not divine, heroism. Jesus can promise the robbers that they will be that day with him in paradise (Lk 23:43), but Sidney Carton can achieve his immortality only in art. However, most of us, I believe, would agree with Woody Allen who said,

> I don't want to achieve immortality through my work, I want to achieve it through not dying (pg. 260). [8]

Unfortunately, the available images of 'not dying' are typically either sketchy, as with the Christian, or grotesque. In *The Facts in the Case of M. Valdemar*, Edgar Allan Poe presents a man mesmerized "in articulo mortis" (pg. 269). [9] The narrator and hypnotist can calculate the hour of expected death because Valdemar suffers from a progressive wasting disease, but in some sense Valdemar in his inevitable mortality is like us all; for, as the inhabitants of Samuel Butler's Erewhon say,

> To be born . . . is a felony – it is a capital crime, for which sentence may be executed at any moment after the commission of the offence (pg. 145). [10]

Poe's story, readable at first as a bizarre science fiction and at second as a flagrant satire, has the time from the narrator's 'conception' of the mesmerizing project to its end equal nine months, the last seven spent with Valdemar somehow suspended by mesmeric intervention. At a key moment in entrancing Valdemar, the narrator says "[I] proceeded without hesitation – exchanging, however, the lateral passes for downward ones, and directing my gaze entirely into the right eye of the sufferer" (pg. 273). This ostentatiously objective rhetoric of science, on second glance, conceals a satire of extreme unction. Indeed, on a subsequent visit, the narrator elicits vibrations from the tongue of the unbreathing, cold Valdemar, and they say, "*I am dead*" (pg. 277). Finally the narrator decides to try awakening his subject. The story ends with this paragraph:

> As I rapidly made the mesmeric passes, amid ejacula-
> tions of "Dead! Dead!" absolutely *bursting* from the
> tongue and not from the lips of the sufferer, his whole
> frame at once – within the space of a single minute, or
> even less, shrunk – crumbled – absolutely rotted away
> beneath my hands. Upon the bed, before that whole
> company, there lay a nearly liquid mass of loathsome
> – of detestable putridity. (pg. 280)

At the most obvious level, this ending suggests that there are some things "that man was not meant to know"; that primal disobedience, such as seeking immortality, may appear to work for a pregnant while, but ultimately the divinely-ordained human dissolution will have its way.

But at a deeper level, this is a grotesque, dirty joke. The ejac-ulations of the tongue parody the ejaculations of a penis and the quick, spasmodic shrinking "beneath my hands" equates unnatural science with masturbation. Instead of describing fertile seed, the story reveals its narrator's own anxieties by ending with "a nearly liquid mass of loathsome [...] putrid-ity." In Genesis, the very instant Adam and Eve ate the apple, "they knew that they were naked" (Gn 3:7). With mortality comes sexuality; those who seek immortality, the power of the gods, seek, perhaps unknowingly, to exchange procreation for creation. Mary Shelley's Frankenstein [11] can restore dead flesh to what may well be permanent life, but the monster, more human than his creator, seeks only a bride, while Victor, like Poe's masturbatory narrator, holds off death with his own hands alone. In *Interview With the Vampire,* Anne Rice's youthful auditor, when he hears the vampire's first description of drinking away someone else's life, says, "'It sounds as if it was like being in love'. The vampire's eyes gleamed. 'That's correct. It is like love', he smiled" (pg. 31). [12] But, of course, it is a love without procreation. Immortality, for the angels, for the devils, and for the creatures of modern science, is a childless state, and to that extent a denial of human potential and of human happiness.

Freud, in *Beyond the Pleasure Principle* [13], suggested that "we have adopted [...] the hypothesis that all living substance is bound to die from internal causes [...] because there is some comfort in it", meaning that all our own failures and our own ultimate demise seem less terrible if seen as either comparatively small or as inevitable. He goes on to assert that "The notion of 'natural death' is quite foreign to primitive races; they attribute every death that occurs among them to the influence of an enemy or of an evil spirit." Freud does not seem to recognize that our seeking of fatal causes – heart failure, cancer, gunshot – reflects no different motive. Instead, in the spirit of Victor Frankenstein, Freud expresses admiration at the writings of August Weismann

> who introduced the division of living substance into mortal and immortal parts. The mortal part is the body in the narrower sense – the 'soma' – which alone is subject to natural death. The germ-cells, on the other hand, are potentially immortal, in so far as they are able, under certain favorable conditions, to develop into a new individual, or, in other words, to surround themselves with a new soma. [13, pg. 616–617]

This is an amazing statement. First, Freud's utter silence here about earlier divisions of the living substance into body and soul reveals a powerful scholarly blindness which can be motivated, one supposes, only by a desperate need to believe

that some progress is being made in the eternal human con-
frontation with death. Second, the focus on the germ-cells,
"on the other hand," is as isolated and masturbatory in its
own way as Poe's focus on mesmerism, another trick of the
mind, like Freud's notion of the death wish, to hold back
the ultimate terror. And third, this notion of immortality for
the germ-cell reduces the human being as we would normally
view it to a mere convenience. While this may be the view
of modern sociobiology observing what Richard Dawkins has
called "the selfish gene" [14], it has little to do with the aspira-
tions of individuals.

But surely we are not our mere bodies. If one lost a finger,
the self would not change. But what if one lost an arm?
Or the ability to procreate? It is clear that we are not much
like our younger selves at the age of, say, three, when we were
all prepubic, utterly dependent, and largely ignorant – indeed,
there may be few atoms in our living bodies that have not
been replaced over the years –yet we like to think of ourselves
as continuous. This is in part an example of the famous philosophical conundrum of the farmer's axe: "Have you had that
axe a long time?" – "Oh, yes. Twenty years. I've replaced the
handle three times and the head twice."

The persistence of the individual is a fantasy, clearly, yet a
productive fantasy without which we would have no sense of
self, and hence without which the very notion of immortality
would be reduced to mere persistence, a state not unlike that
of a rock.

Modern science fiction has, of course, imagined selves con-
cretized if not in rocks then in silicon. In Clarke's *The City
and the Stars*, citizens of Diaspor live so mind-numbingly long
that they eventually voluntarily walk back into the "Hall of
Creation" where machines "analyze and store the information
that would define any specific human being" [15, pg.15] and
then they give themselves back up to silence – one shouldn't

call it death – until recalled by the Central Computer at some random future time to live with a newly randomized mix of ten million of Diaspor's billion potential citizens. Yet in this immortal utopia, where merely to speak the name of desire is to have it materialize, our hero Alvin is not just another revenant but "in literal truth [...] the first child to be born on Earth for at least ten million years" [15, pg. 17]. It is he who brings fecundity and progress back to a stagnant world. There is no real human life without mortality, without the risk of death. From among all the traits that characterize us, we choose to call ourselves "mortals". This is the wisdom of Pinocchio.

In William Gibson's *Neuromancer*, one character is a so-called "construct", a computer chip containing the knowledge and personality of a famous denizen of "cyberspace," the virtual reality of the infosphere. He is activated by some "meat" characters who need his help, and he agrees to aid them but with one proviso: at the end of the adventure, "I want to be erased" [16, pg. 206]. Apparently disembodied immortality is as much a trap for Dixie Flatline as aging, embodied mortality is for Tithonus. We understand why, I think, when Case, the protagonist, tells Dixie that "'Sometimes you repeat yourself, man.'" "'It's my nature,'" Dixie punningly replies [16, pg. 132]. Given enough time, and no body to respond to a changing environment, we would all repeat ourselves, living out patterns, no matter how grand, that lead ultimately to the merest repetition, and hence the destruction of any sense of individuality.

Thus it is that the sentient computer HAL, in Clarke's *2001: A Space Odyssey*, ceases to be a character – an individual – but continues to function as a computer when his "higher function" boards are removed and he is reduced to repeating the calculations and self-identifying serial numbers first programmed into him. [18 pg. 156–157]

It is for the same reason that Olaf Stapledon in *Star Maker* praises not swarming "hive minds" that obliterate the individual but the "intricate symbiosis" [19, pg. 255] represented by a perfect marriage, by that "prized atom of community" [19 pg. 257] in which two may depend upon each other – and procreate – but in which each maintains essential individuality, and risks individual death.

Against this view, we have *Blood Music*, in which Greg Bear lets loose a plague of "intelligent leukocytes" and the world is transformed, all of us ultimately parts of a planetary hive mind. The protagonist says, "if I die here, now, there's hundreds of others tuned in to me, ready to become me, and I don't die at all. I just lose this particular me [...] it becomes impossible to die" [20 pg. 197]. Bear's protagonist may believe that, but identical twins do not: no matter the duplication of information in another copy, the death of the individual as contemplated by that individual is death indeed. And the capacity to die is a great, self-defining freedom, the ultimate existential freedom according to Sartre and Camus and the very ground of conflict between the individual and the state, as seen in the hospitalized, limbless combat victim in Dalton Trumbo's *Johnny Got His Gun* [21], in Brian Clark's tube-fed paraplegic in *Whose Life Is It Anyway?* [22], and in D-503, the protagonist of Eugene Zamiatin's *We*, after the "splinter [of imagination] has been taken out of [his] head" and he is reduced to a permanent, idiot grin, for "Reason must prevail" [23, pg. 217–218]. This happy state of inevitable obedience is the ultimate Eden, and the splinter removed from D-503 is the "thorn" of the plant Gilgamesh sought, its prickle reminding us that we are alive as individuals only when we are subject to death.

It is said that when Michelangelo completed the idealized Medici tombs ordered by Pope Clement VII someone remarked on an absence of realism. "Who will care," the great sculptor replied, "in a thousand years' time, whether these are their features or not?" [24, pg. 399]. Indeed. On the day jazz great Duke Ellington died, John Chancellor began his nightly television newscast by saying that "Edward Kennedy 'Duke' Ellington died this morning of cancer of the lungs and pneumonia. Later in the program we'll hear him play for us" (pg. 76). [25] Idealized in stone or vinyl, the great achieve immortality not in themselves but only in their leavings, an immortality that supplants, and hence defeats, the self.

St. Paul promises us that here on Earth

> we see through a glass, darkly; but then [after Judgment Day] face to face: now I know in part; but then shall I know even as I am known (1 Cor 13:12).

This notion of ideal knowledge in eternity is not limited to the Western world. The voice in Brihadaranyaka Upanishad, pleads:

> Lead me from the unreal to the real! / Lead me from darkness to light! / Lead me from death to immortality! (Bartlett 56:20) [3]

But who is this *me*? Who is this *I*? When Moses asks on Mt.Sinai to see God face to face, God, who favors Moses, withholds this favor "for there shall no man see me, and live" (Gn 34:20). St. Paul understood this, too. Speaking of the resurrection after Judgment Day, he says

> Behold, I shew you a mystery; we shall not all sleep, but we shall all be changed. In a moment, in the twinkling of an eye, at the last trump: for the trumpet shall sound, and the dead shall be raised incorruptible, and we shall be changed. For this corruptible must put on incorruption, and this mortal must put on immortality (1 Cor 15:51–53).

When we put on incorruption, we are all changed: we are changed into ideals, into endless repetitions, into sterile vampires, childless angels, works of art, computer chips. We are changed into objects for the contemplation of others but in the process we lose our very selves. Immortality is a self-defeating fantasy, a desperate defense against death. Finally, who would choose such a neutered eternity? Not Tennyson's *Tithonus*:

> Let me go; take back thy gift. / Why should a man desire in any way / To vary from the kindly race of men, / Or pass beyond the goal of ordinance / Where all should pause, as is most meet for all? Release me, and restore me to the ground. [17]

References

1) Anonymous; *The Epic of Gilgamesh*; Trans: Sandars, NK; Penguin, 1960

2) Calvino, Italo; *If on a winter's night a traveler*; Trans. William Weaver; Harcourt, 1979

3) Bartlett, John; *Familiar Quotations*; (1980, 15th ed.); Little, Brown

4) Shaw, George B; *Man and Superman* (1903); Penguin, 1969

5) Milton, John; *Paradise Lost* (1667); Editor Merritt Y. Hughes; Odyssey, 1962

6) Clarke, Arthur C; *Childhood's End* (1953); Ballantine, 1972

7) Dickens, Charles; *A Tale of Two Cities* (1859) T. B. Peterson

8) Peter, Laurence ; J. *Peter's Quotations: Ideas For Our Time* (1977); Bantam

9) Poe, Edgar Allan; "The Facts in the Case of M. Valdemar." 1845. Ed. Philip Van Doren Stern. *The Portable Poe.* New York: Viking, 1973

10) Butler, Samuel; *Erewhon* (1872), New York: NAL, 1960

11) Shelley, Mary; *Frankenstein* (1818), Editor: Joseph, M.K. Oxford University Press, 1969

12) Rice, Anne; *Interview With the Vampire* (1976); Ballantine

13) Freud, Sigmund; *Beyond the Pleasure Principle* (1920) in *The Freud Reader* (Editor: Gay, Peter); Norton, 1989

14) Dawkins, Richard; *The Selfish Gene* (1976); Oxford University Press

15) Clarke, Arthur C; *The City and the Stars* (1956) Harbrace; pg.15

16) Gibson, William; *Neuromancer* (1984) Ace; pg. 18

17) Tennyson, Alfred; "Tithonus" (1860) in: *Selected Poetry* (1956); Editor: McLuhan H.; Holt

18) Clarke, Arthur C; *2001: A Space Odyssey* (1968) New York: NAL

19) Stapledon, Olaf; *Star Maker* (1937); Dover Puplications Inc.

20) Bear, Greg; *Blood Music*; New York, Ace, 1985

21) Trumbo, Dalton; *Johnny Got His Gun* (1939) Bantam, 1970

22) Clark, Brian; *Whose Life Is It Anyway?* New York; Dodd, Mead, 1978

23) Zamiatin, Eugene; *We* (1920). Trans: Zilboorg G; Dutton, 1952

24) Fadiman, Clifton, ed. *The Little, Brown Book of Anecdotes* (1985) Little, Brown

25) Rabkin, Eric S; *The Fantastic in Literature* (1976) Princeton, New Jersey; Princeton U Pr, 1976

Timeconsciousness in Very Long Life

Manfred Clynes, Ph.D.

Counting is not time – there is no time to count.

In the following essay, I will leave immortality to the Good Lord, and will try to be absolved of some hubris by dealing with individual life of only a few million years long. I hope not to disappoint readers through this, at least not for their first million years.

What is time? We have been all too much influenced by physicists, who have described it as a dimension: an infinitely thin straight line, or somewhat curved if you consider Einstein, along which events move. What happens is that t moves from t1 to t2, two points along that line – the beginning and the end of the event. Time as an infinitely small point goes from its place at t1 to the place t2. And we additionally have been brainwashed to consider it going from left to right. Under quantum level conditions, at the scale of Planck's constant, time may even reverse for very short instants. To ask "how fast does it move along that line?" is a meaningless question for the physicist. Yet the relative rate at which time goes depends on the coordinate system; the frame of reference.

What is missing from this view is the *present*. In physics as in human life, time converts potentiality to actuality. Einstein was uncomfortable too in banishing the present from his theory (or not encompassing it). But I have yet to find a physicist today who misses the present, as a physicist. It seems simply nothing to worry about. But, interestingly, the present is all that exists for us as humans. The past is gone and the future is not here yet. What we experience appears to be a continuing present, as long as we are conscious – also memories of the past, and anticipations of the future are experienced in the present. The present is always with us. We can say, even, that the present itself is unchanging and only its contents change. In a sense, the present is eternal. As long as you live. What is it then that we have that creates the present in us?

We call it timeconsciousness. In the place of an infinitesimal point sweeping along the time line, we have a finite time slider on which we sit, so to speak, that slides along time, carrying us piggyback. The slider is the omnipresent present. In my earlier work of the seventies I showed that a good measure of the duration of the present is about 180 milliseconds. It is the duration of a syllable, of the minimum time during which a decision cannot be reversed, motor reaction time, and the time for which a slowly moving object is seen as moving, rather than stationary. More recently we have also found that it is the preferred time for a composer's fastest independent pulse components, at least in Mozart and Beethoven Allegros.

But a word is not a substitute for understanding, or should not be. We know little about timeconsciousness. Through our work in music, we have found evidence for four different clocks and processes involved in different aspects of timeconsciousness in our brain. These clocks involved in music operate in our timeconsciousness. Since it is possible to think music while dreaming, they are transferred also to our changed timeconsciousness while dreaming – some of them. Somehow, in dreaming our timeconsciousness is rescaled – how come? We don't know.

As we readily surmise today, different animals (conscious machines) have different timeconsciousnesses. There is nothing absolute about our timeconsciousness. On a different galaxy, say, a living being could exist to whom night and day would be a flicker. Our timeconsciousness is purely relative to our being human.

TIMECONSCIOUSNESS SCALING

This invites the consideration of scaling of timeconsciousness. We will learn it from how the DNA does it. And redesigning ourselves for long life we can take advantage of variable timeconsciousness scaling.

What is the timeconsciousness scaling of a computer? Or, what is the timescaling of a computer? The computer's idea of time is that it has no idea of it at all. All it knows is a series of numbers, the time stamps. What the time interval is between these numbers is entirely arbitrary. We can increase the duration of a computer's tick (computing cycle) and the computer would not know: all its answers would be the same. Any calculation it can do at any tick size within its technologic ability will provide the same answer. A string of ones and zeros cannot give either it or us the experience of time. A flaw in the Turing test is that it leaves out time. And as long as we are modeling ourselves along the lines of a computer as we understand it today, we will have no timeconsciousness at all. And that means we would not be conscious.

Zeros and ones, numbers, cannot replace the uniqueness of time for us. Indeed it can in a four-dimensional matrix, in Minkowski's representation, and for Einstein it requires an imaginary axis to distinguish it from spatial dimensions. But numbers cannot tell us about the experience of time, nor for that matter of space. Our brain and our nervous system are designed so as to distinguish time and space. It gives time attributes that cannot be represented by a simple imaginary operator. Nor do neural pulse trains in our brain distinguish between space and time. On the other hand, music provides a prime laboratory to examine our relationship to time.

TIMECONSCIOUSNESS AND THE EXTEND LIFE

In considering greatly extended life we need to reflect on the following (these points cover only a few aspects of concern related to time; we shall here leave out social concerns):

Until timeconsciousness is understood sufficiently to enable computers or robots to have it, and until they become conscious, extending life beyond the limits imposed on the materials from which we are made today is not possible. Replacing biologic materials, homeostasis, metabolism, human reproduction and of course memory and thinking (as we know it today, insofar as we know it at all) with more stable non-biologic materials, nanotechnology, and biostructural design will suffice only once we know how consciousness arises. I am convinced that consciousness is not a function of complexity in itself, nor that a very high degree of complexity is a prerequisite for consciousness. It is something else.

Once we will know how to create consciousness, we are in a totally different world compared to which cloning will be mere child's play. It may be only hundreds of years before that happens. It could be thousands. No one really has a clue today. To me a salient thing about consciousness is that the more one removes its contents, the closer one gets to 'pure consciousness' i.e., contentless consciousness. Helen Keller was not less conscious than most of us, but probably more so. Various sensory inputs interfere relatively little with one another in consciousness, we can hear, see, smell, touch simultaneously and more. If we create conscious machines, (which are what we tend to think we are today), but of different structures than those we know, then there is a real possibility of dissolving the ties to time that we are accustomed to experience.

Choosing your Own Timeconsciousness

When creating conscious machines, we would have the freedom to design ourselves to have different timeconsciousness rates, and also variable timeconsciousness rates to suit our needs. What will our needs be? In terms of travel, we can envisage much: for space travel, slow down our time consciousness, say by a factor of 10,000. What's 10,000 years if they pass like one year or even one month? And if you combine this with relativistic Einsteinian time slowing, you have some favorable space travel conditions. If you have millions of years to live, timeconsiousness contraction does not matter; there will be plenty of experience left. It will also be useful to have contact with other forms of life, at least in our galaxy and adjust your time-consciousness to other beings.

More likely, often, for other than space travel, we would use the opposite, an expanding time consciousness, a speeding-up adjustment using nanotechnology, or picotechnology; thinking could be say 10,000 times faster than we are used to. What would happen then? A year would last 10,000 years. The seasons would not change for 2,500 years. The factor of 10,000 is perhaps a somewhat extreme example here; you could adjust your time consciousness rate as you wish, maybe sometimes only double it or triple it, depending on the situation you are in.

Aging will be eliminated as we know it. Aging would turn into a rather serious problem of memory accumulation however. All the matter in the universe is insufficient memory to store the state of all matter in the universe. As the age of individuals mounts into millions of years relatively more and more matter will be needed to store the memories. Eventually, Freud notwithstanding, a selective erasing of memories may need to take place – a hot political issue of the very distant future.

More serious would be the problems posed by our emotions and their communication. As I have shown, the expression and generation of emotion by communication such as gesture, dance, and music is highly time dependant. Time forms, which we call sentic forms, form the vocabulary of our inherent language of emotion communication and generation. Shape (in time) is quality. What happens to those under time-consciousness transformation?

Take laughing as an example. Can we laugh in our dreams, on a different time scale, and yet experience laughter? A fast thinking person still laughs at similar rates of ha's than a slow thinker, even Marvin Minsky. The difficulty lies in the mysterious so far unexplainable inherent connection between the time form and the quality of feeling that it embodies. Speed up timeconsciousness by a factor of 10,000 and laughter would have to be speeded up in a similar ratio. Simply as a motor control problem, even if we are greatly downsized, this appears to be unrealizable (atoms within our brains do not now move many thousands of times faster than our arms). The inertia of matter, which we cannot remove, will conflict with the speed of thought, which in essence is free from material considerations of inertia, that is the speed of computation vs. the speed of motion: the speed of electrons and of photons versus that of protons and neutrons – leptons vs. hadrons. The duration of orgasm seems to be more fixed in us than the duration of thinking. The inherent time-link of a sigh, of a caress, of a joyous leap all seem part of our nature we would from our perspective want to keep qualitatively, or improve if possible, under timeconsciousness transformation.

Obviously, emotional communication and generation, caresses, sighs, would have to become free of – (present) body constraints. Can that be possible without changing its quality? Or, would one hope for a transformation of quality into something desirable unknown? It may seem unlikely that the exploration of changing qualities of the range of experience under timeconsciousness transformation would yield new qualities preferable to those we know. But we do not know. We do know already that qualities of experience in dreams are rather similar to those in the waking state, even during this limited transformation. How far can this be extended? What new qualities can be created by more systematic and extensive timeconsciousness transformations?

Already we sense that music is too slow to give us its feeling and content: the visual sense, video, is overtaking it in our culture, through its more concentrated meaning transfer, with its parallel input instead of serial. A symphony takes too long for our relatively fast paced world. Music is used as background for the visual. What will happen to music under timeconsciousness transformation? Its art will also be transformed. Problems and related aspects of that are considered next: logogenesis.

LOGOGENESIS

Logogenesis is the 'invention of nature' which appears in evolution that substitutes a quality of feeling, of experience, for convoluted thought and for reflex activity. Logogenesis is what creates qualia, the flavor of life. Like morphogenesis it is genetically controlled. Quality and its cognitive substrates arise in our brains through a genetic program details of which we do not yet know. For example, the feeling of laughter, its funniness, is indissolubly linked to the pattern of expression. Rednesss or sweetness is indissolubly linked to a particular pattern of neural and synaptic activity, which in itself looks similar to other neural activities and in no way discloses the quality of redness or sweetness. Yet these ('unlearned') qualities of experience remain unchanged throughout a lifetime, and often even until the last moments of life. We do not know what controls logogenesis. The earliest ingenious invention perhaps is the substitution of the feeling hunger for other ways to regulate food intake. That feeling tells us when to eat, what to eat, and how much to eat and urges us to move to find food. What an amazing thing! How would we invent something like this?

But perhaps even more ingenious is the sexual feeling. With it the chances of reproduction (say of a mammal) increase from infinitesimal to a viable, even likely number. That feeling in its manifestations makes the continuation of the species, and evolution itself, possible.

What invented that feeling and placed it in its environment so it could, much of the time, function? Is it itself evolving? Clearly it is genetically programmed. If the sexual organs are generated through morphogenesis, the feeling, conscious and subconscious, is generated by logogenesis. So is the intense feeling of orgasm, different from other feelings. All qualia are created by logogenesis. We do not know yet how that works, and how it will change under timeconsciousness transformation. But a computer or a robot will need it to feel. It does not seem that zeros and ones are a way to obtain its function. How then? We have a terrible way to sweep under the carpet anything we do not understand and pretend it does not exist. It is time to see that the emperor has no clothes. Neither consciousness, nor timeconsciousness is possible for a machine until logogenesis is understood. It is no good to say that we are software, and can be up- or downloaded. Brain processes and experiences are both analog and digital, continuous, and discrete. Today, you can make a computer laugh but it will not enjoy it; even if it says it does. It is time to stop deceiving ourselves. Turing deserves better than that.

My computer plays beautiful Beethoven. I am happy that it does. But it is not. Nor will it be until it understands Beethoven. Without timeconsciousness that is impossible. Without logogenesis that is impossible. But give it time, maybe some day it will join us in all this. Then it will become one of us and we will become one of them. The Good Lord probably (or improbably) knows when and how this will happen. But I would caution us to try to find the probability. A little faith is probably programmed into our logogenesis system.

Related References

Abeles, M; "Time is precious, Perspectives", in *Science,* 304, pg. 523–524, 2004

Barnes, J et al; "Reqirement of Mammalian Timeless for Circadian Rhythmicity" in *Science* 302, pg. 439–445 2003

Clynes, M; "The Future Compassionate Computer": presented at MIT Medialab Celebration Oct 1999, 2000, www.superconductor.com

Clynes, M; "Entities and brain organization: Logogenesis of meaningful time-forms", in *Proc. of the Second Appalachian Conf. on Behavioral Neurodynamics.* Origins: brain and self organization.ed. K.H. Pribram, Erlbaum Press, N.J. pg. 604–632, 1994

Clynes, M; "Time-forms, nature's generators and communicators of emotion", Proc. IEEE Int. Workshop on Robot and Human Communication, Tokyo Sept. 1992, pg. 18–31

Clynes, M; "Time, Timeconsciousness and Music" ; Proceedings of the First Internatonional Conference on Music Perception and Cognition, Kyoto, Japan, 1989, pg. 124–130

Clynes, M; "Methodology in sentographic measurement of motor expression of emotion" in *Perceptual and Motor Skills*, 68, pg. 779–783, 1989

Clynes, M, & Walker, J; "Music as Time's Measure" in *Music Perception*, Vol.4, No.1, pg. 85–120, 1986

Clynes, M; "Specific human emotions are psychobiologic entities: Psychobiologic coherence between emotion and its dynamic expression", Commentary, in *Behavioral and Brain Sciences*, 1982. No. 3. pg. 424–425

Clynes, M; "The communication of emotion: theory of sentics", in *Emotion: Theory, Research and Experience, Vol. 1 Theories of Emotion*, R. Plutchik, H. Kellerman (eds.), pg. 271–300, Academic Press, New York, 1980

Coull, J et al; "Functional Anatomy of the Attentional Modulation of Time Estimation" in *Science*, 303, pg. 1506–1508, 2004

Crick, F; *The Astonishing Hypothesis,* Touchstone Books, New York, 1995

Damasio, A; *The feeling of what happens, body and emotion in the making of consciousness,* Harcourt, New York, pg. 386, 1999

Dennett, DC; *Consciousness Explained,* Little brown, Boston, 1991

Feynman, R; "Thereis plenty of room at the bottom" http://nano.xerox.com/nanotech/feynman.html

Ikegaya. Y et al; "Synfire Chains and Cortical Songs: Temporal Modules of Cortical Activity", in *Science,* 304, pg. 559-564, 2004

Kurzweil, R; *The Age of Spiritual Machines,* Penguin Putnam, New York, 2000

LeDoux, J; *Synaptic Self,* Viking Penguin, New York, pg. 406, 2002

Minsky, M; *The Society of Mind,* Simon and Schuster, New York, 1985

CONFESSIONS OF A PROSELYTIZING IMMORTALIST

Shannon Vyff

Why Immortality? Isn't that selfish? Doesn't God, fate, and evolution tell us when to die? I hear this all the time, shortly after I meet people at church, on vacation, in buses, in line at the grocery store, at parties, and at family gatherings. It all begins like a perfectly normal conversation with questions of the weather, or what the person has been doing for the day. I usually mention my schoolwork explaining that I started studying nutrition after my 85-pound weight loss (always a motivating, intriguing story for others). Ears perked, they want to know how I did it. I tell of how I got involved in Calorie Restriction (CR) with *Optimal Nutrition* online mailing list. [1]

Through the CR Society, I've had fun being in Oprah Magazine, in Marie Claire, on 20/20, traveling to the CR conference to hear the latest research on this anti-aging diet, and meeting interesting people who use themselves as human guinea pigs. I myself have sent in blood work and various other test results keeping track of my bio-chemical markers of aging to help out with the human studies. People say I look great, (especially if it is someone I have not seen for several years and they remember me as being much larger), but then they naturally want to know why I started on this diet.

I start by telling about how, throughout high school, I followed the biosphere dome and Dr. Walford, [2] because of my love of science fiction. While growing up, I always had the dream of traveling to the stars. I guess I was hoping that in my 20's there would be self-contained spaceships heading to colonize Mars, and that I could sign up. When the biosphere was in the media usually there was also mention of the calorie restriction (CR) diet so it was always in the back of my mind as a slimming and anti-aging diet.

In my early twenties I choose to start a family, and learned from La Leche League [3] about the healthiest way to do that. When it came time to wean them from the breast milk that gave them the thickest neuron connections, quickest reflexes and strongest immune systems that with our current technology they could have, I had to look at what real food was best to put into their growing bodies. I lived in Eugene, Oregon where there was so much support – with actual organic-only restaurants, free health newsletters, and a high population of health conscious people – that I was introduced to a new way of looking at food. I realized that we replace many cells every 6 months and most of our body every seven years, so that we truly become what we eat many times during our lifetime.

I also started learning about fungicides, germicides, pesticides, rodenticides, herbicides, antimicrobials and how they collect in body tissues. I read studies such as the recent one funded by the National Institute for Environmental and Health Sciences [4], that looked at pre-school age children in Seattle and found that organically fed children vs. non organically fed children had six times less organophosphorus (due to pesticide exposure) in their urine.

Though I had started eating more natural and organic foods, I was still 205 pounds three months after the birth of my last child, and I started thinking I needed a real diet! I began reading about all the diets out there and remembered that life extension diet I'd heard of back in high school. That is how I picked up the book *Beyond the 120 Year Diet* by Roy Walford, M.D. when my third baby was four months old. As I was still breastfeeding and still had the post-pregnancy slimming hormones, I did lose weight more rapidly than is recommended for CR. I lost 85 pounds in six months, but a much more gradual decrease is recommended. One should consider losing large amounts of weight (perhaps over 40+ pounds) over 2–4+ years or consider losing weight no faster than a pound per month. This is because the release of toxins stored in the bodies fat tissues and the stress of the weight loss on the system could foil extending life span in one's final years. [5] When I was beginning the diet I used software to help me balance what nutrients I needed. I started making mega-muffins for my family to eat (like human lab chow with organic chocolate chips thrown in for my kids to eat them). [6] I found that as my weight came off I had more energy and clearer skin.

Usually, by now, people want to know why I want to follow what they see as an extreme diet. Saying that it is the only scientifically proven way to extend my life [7], hormones, anti-oxidants, exercise or high carb/low carb diets – generally are not enough. Usually people express that I must have iron will power and torture myself with hunger, something they could never do. I assure them that for me it is easy; there is great support from the CR Society. People do it many different ways. I recommend *The Anti-Aging Plan* by Roy & Lisa Walford for an easy to read beginners introduction, though *Beyond the 120 Year Diet* [5] has better detail & more current science references.

As I became enamored with the science of CR, and embarked on it as a way of life, I started seeing more references to it in popular culture. [8] The ongoing experiments in primate & human studies also made it very real for me. The National

Institute on Aging has been conducting a long-term study, since 1987, of CR on rhesus monkeys. In 1999, the NIA researchers stated:

> Emerging data from studies of CR in rhesus monkeys show promise that the model is working in a manner similar to that seen in rodents thereby strengthening the possibility that the well-known effects of CR on life span, disease, and aging processes may be generalized to all species. [10]

Regarding the NIA study, *Modern Maturity* states "The incidence of diabetes (...) is greatly reduced in monkeys on a restricted diet. The monkeys also show fewer signs of spinal arthritis, a common condition they share with humans." [9] These monkeys show other signs of reduced aging, such as a prevention of age-associated decline in melatonin levels. [10;11] When I was at the CR conference I was able to see a presentation by the researchers working with the rhesus monkeys and hear about how they were doing, even how they were enjoying their new living environment! Seeing the drastic difference between the ad-lib monkeys and the calorie restricted ones made it all the more real for me.

On the 'easiness' of CR, I explain that it is important for me to see as much of life as I can. Therefore, I have transferred my comfort foods into more healthy ones, like air popped popcorn, raw veggies, or green tea. This helps me reach the long-term goal I always have in my mind: to keep my brain sharp as long as possible. CR is a way to help me reach my other goals. I want to stay alive to a time when we change the ratio of expenditure from 400 billion US dollars to the military [12] with only approximately three million to aging research [13] to something more appropriate to combat and eradicate the greatest killer of all time – aging. There are lots of figures out there, and in all of them America's military spending dwarfs any other countries and even many countries entire GNP. In discussions, when some are particularly pessimistic about the ability of our current society to change, I point out that I notice more people every year are committed to bringing about complete political reprioritization. This makes me optimistic. Yet when I look at the past 2000 years of written human history, I know the changes I envision, including immortality, may not occur within my lifetime, even with CR. So I have a back up plan:

This is when I show my braided gold medic-alert while I say I'm signed up with ALCOR to be frozen when I die (actually 'vitrified', a new technique with virtually no cellular damage). [14] To me, it beats being buried in the ground or becoming ashes. I then explain why I hope to see the future: I believe we will soon (100 years or so) learn to not only end aging but to reverse it and have handy things like brain back-ups if we have accidents, effectively making us immortal.

This is where the various conversations with strangers, friends, and family from all walks of life, in all sorts of places, really get interesting! People understand eating healthy; they appreciate hearing practical things to help them now in 2004. It is easy to have someone believe you when you talk of things that can help right now, but when projecting into the realm of science fiction, or just humanity's future, I run up against a wall. Facial expressions change and I'm asked why I'd want to live forever or believe on blind faith that it is possible. This is where I hear that God, evolution, or fate knows when we are to die and that it is selfish to extend our lives unnaturally. I point out that a primitive society life span is around 25 years of age with 40 being a rare old man. Being old is in fact not natural in nature's setting. Living 50–60+ years for a significant percentage of human populations is a modern adaptation of the last 100 years of human history. The majority of deaths also changed from those of young children to those of old people. Our scientific and technological advancements in the past 100 years are already giving us vastly extended life spans [15–17].

This brings me to my counter point to the 'selfish' accusation. The basis of my argument is that extending life through medical interventions helps us to do more of what God, fate, or evolution has planned for us to do. This is why we are given, or have developed, the intelligence to stay alive longer. When we stay healthy longer through CR, or anything that we develop in the future, we can work longer and can give more to the spiritual, scientific, or non-profit organizations we affiliate with. I for one would love to be able to donate more to the Methuselah Mouse Prize for anti-aging research [18]. If the award for that was more near the largest ever US lottery jackpot of 363 million, rather than the fifty thousand dollars currently donated, we would shortly end aging. With more prize money there would be a lot more contenders for the prize that will be awarded to whoever can significantly reverse aging in a mouse, or postpone it.

With a longer life, many opportunities would open up, for new careers, new travel, and exploring the ever-expanding questions of the universe. These conversations with people can be very inspirational. It is surprising how creative people can be when you open them up to talking about the future.

At times I am dismayed that with our current technology my life will be so short. Yet my heart revels with what epiphanies I inspire in my children about how things in this world interact. I get a rush when I explain to them in simple terms (and they actually get it) – about why it is important for our country to balance the budget, how to support universal health care, how people live in oppression even today, or just explain any little piece of the puzzle to them. With the help of the Unitarian Universalist Church's education programs, I teach them about the world's religions. What other people believe, their histories and how their cultures (and even our own) are affected today by their religions. In class they also learn about community service, about self-esteem and how caring committed people have affected change. I see my children struggling to balance their emerging ideals with what is 'cool' as seen in our popular media culture versus what their own hearts tell them. It can be overwhelming for anyone young or old to hear of all the wrongs to be righted. To my children I like to mention that if the Earth's 4.5 billion year history were represented as 24 hours, the 2000 years of written history would be a mere second or two, and in that brief time we humans have been on an increased path of happiness and wisdom.

In all these wonderful, almost daily, deep, philosophical discussions I get into with various people and my own children about the nature of this universe, I mainly try to instill hope. I think the ripples I create might spread. It's not just about proselytizing (although I'd love all to join the immortalist cause). It is about the little sound bites I give people (like the eye-opening things you will no doubt read in this book) such as neural chip implants for rats, photographic memory for fruit flies, and cat brains that have been frozen and then brought back to normal looking electrical activity – things we have actually been able to do. I like to think these sound bites will come out elsewhere in other people's conversations, and they may be inspired to learn more. In this process of becoming more curious and aware, they internalize how they can affect change.

When younger people today (and older people who have already accumulated more wealth and power) think about such things as being aware of how research money is spent and their own taxpayer money is used, they can vote in to office the people who will spend it most effectively for them. I am thankful that the Immortality Institute has been created to help change how some of this wealth is spent. Some day we could even be supporting our own Immortalist candidates for office, supporting foreign aid, increasing science and technology fields, and ending aging! I know most of the people I've met who are already immortalists have not gotten into the movement until they were a young person in an old person's body wondering what happened. This is why I talk to so many people to try to spread awareness in the younger generation. It is why my children are being raised with the naturalness of transhumanism (being open to what we may become).

If you ask my brilliant seven-year-old what she thinks about immortality she will boldly start talking about what things she could do with a robot body. My highly imaginative four-year-old son will say how he wants to be a scientist that discovers how to end aging since he never wants to die! My angelic two year old hugs and kisses everyone, and reminds me of the basic instinct to feel pleasure by helping others, as she shares her food and toys, along with her vibrant spirit of life.

This book is exhilarating in its scope, and in its predictions. The best way to predict the future is to help create it. I hope that things you read in this book will touch that spark of optimism each of us has at the emerging of our own consciousness, as we rush through our childhood eager to understand our place in this universe.

References

1) http://www.calorierestriction.org

2) http://walford.com/biosphere.htm

3) http://www.lalecheleague.org

4) Curl, Cynthia L & Fenske, Richard A & Elgethun, Kai; "Organophosphorus Pesticide Exposure of Urban and Suburban Preschool Children with Organic and Conventional Diets?" in: *National Institute for Environmental and Health Sciences Journal* (2003, Vol. 111, pg. 3)

5) Walford, Roy; *Beyond the 120 Year Diet*, (2000); pg. 78–80

6) A great online site for tracking nutrition data & diet is http://nutritiondata.com. Some CR recipes including mega-muffins are online here: http://recipes.calorierestriction.org

7) Original study that discovered that caloric restriction extends animal lifespan: McCay CM, et al. ; "The effect of retarded growth upon the length of life span and upon the ultimate body size" in *Journal of Nutrition*, (1935, 10(1)) pg. 63–79

8) Taubes, G; "The Famine of Youth" in *Scientific American* (June 2000)

9) Warshofsky F; "The Methuselah Factor" in *Modern Maturity* (1999, November-December)

10) Roth GS; "Dietary caloric restriction prevents the age-related decline in plasma melatonin levels of rhesus monkeys" in *Journal of Clinical Endocrinology & Metabolism*, (2001, July, Vol. 86(7)), pg. 3292–5

11) For those who are interested in the most current (technical & reference) information on the science of CR, I recommend: Edward J. Masoro; *Caloric Restriction: A Key to Understanding and Modulating Aging* (2002) Elsevier Health Sciences

12) Hellman, Christopher; *The Center for Defense Information's FY2004 Discretionary Budget,* http://www.cdi.org/budget/2004/discretionary.cfm

13) Ellis, Joseph; *The proposed NIA FY2004 Budget and NIA salary information;* http://www.nia.nih.gov/fy2004_congress/ftes.htm

14) http://www.alcor.org

15) Olshansky, Jay S & Carnes, Bruce A; *The Quest For Immortality: Science at the Frontiers of Aging* (2001), Norton & Company

16) Perry, Michael R; *Forever For All: Moral Philosophy, Cryonics, and the Scientific Prospects for Immortality;* (2000) Universal Publishers

17) Bova, Ben; *Immortality: How Science is Extending Your Life Span and Changing the World* (2000); Avon books

18) Various; The Methuselah Foundation: Longitude Prize (2004) Society; www.methuselahmouse.org

Some Problems with Immortalism

Ben Best

In the Land of Oden / There Stands a Mountain / A
Thousand Miles in the Air. / Once every Million Years
/ A Little Bird comes Winging / To Sharpen its Beak
on that Mountain. / And when that Mountain / Is just
a Valley / This to Eternity shall be... / One Single Day.

I heard this English translation of an Austrian folksong during
my second year of university, and it still strikes a deep emotional resonance within me. I have craved to live for eons since
I was a small child, and evocations of the expanses of time
draw me with a hypnotic power. Long before I heard of cryonics I had a rich fantasy life and I would imagine myself happily surviving alone after the rest of Mankind had passed from
the scene and the planet Earth had been turned to volcanoes
and fire. But, although I place no limits upon how long I want
to live, I believe that there are good reasons for believing that
immortalism is an unrealistic goal – and even a self-defeating
goal. 'Forever' is not just a long time; it is eternity and therefore beyond realistic conception.

There are mathematical models that can be used to calculate
valuations in infinite time: I would rather be given $1 today
than $1 in one year's time. Similarly, the value of being alive

for the next year is more important for me than the value of being alive the following year, and much more important than the value of being alive for one year 100 years in the future. The present value of money can be compared with the future value of money in a choice of present money or future money – a mutually exclusive choice. A mutually exclusive choice cannot be made with life because being alive in the future requires being alive in the present – all the more reason for placing greater value on life in the present; but diminishing returns can and do still apply.

Let's say that being alive in 2005 is 98 % as important to me as being alive in 2004. Then I can calculate the value of immortality for me as:

the sum of N from zero to infinity for

$$(0.98)N=1 \div (1-0.98)=50.$$

That is, I value immortality 50 times as much as I value being alive for another year. If this seems unreasonable, then ask yourself: "Is being alive for one year at the age of 100 really as important to me as being alive for one year at the age of one thousand or one million?" It must be that events far in the future, even being alive, must be of less personal significance or urgency than events in the present.

To put the argument in the most forceful terms, if you knew for a certainty that you were going to be obliterated without hope of further life at the age of one million years, would that be significantly more tragic than an age of ten million? Ten billion? Ten trillion? Even after ten quadrillion years, you can never know that you have achieved immortality – that would take *eternity*.

Imagine that you have a million dollars and that you are to allocate that money entirely toward ensuring your survival in any given year. If you allocate the entire million to surviving only for the coming year, chances are very good that you will survive the entire year, but maybe not so good that you will survive the following year. If you allocate one dollar per year to survive one million years you may have a hard time lasting even one year with only a dollar for food, shelter, medicine, self-defense, etc. By allocating more money to the next few decades you increase the probability that you will be able to acquire money to survive in future decades. The analogy with money should be expanded to effort and attention.

I have known an 'immortalist' who argued very strenuously with me when I suggested that physical immortality is probably not possible or deserving of attention. Yet this person would not make the effort to complete cryonics paperwork – preferring philosophical questions. Although many defenders of physical immortality have indeed completed their paperwork, I still think the emphasis on physical immortality is misplaced. It is too easy to step into an open manhole by spending too much time with one's head in the clouds. Even if one's own immediate survival is not imperiled, there is a danger to others – and ultimately to oneself – of distracting attention from real and tractable problems in favor of futuristic and fanciful ones (which may be more entertaining).

Concerning priorities, I think that making cryonics arrangements – and taking steps to ensure that those arrangements are implemented – is good first aid. It could be valuable to work on means of extending maximum life span, but only after adequate attention has been given to preventing death by cancer, cardiovascular disease and fatal accident. Too much focus on the former while ignoring the latter is another instance of having one's head in the clouds. A very sad example in this regard is Frank Cole, who studied anti-aging medicine, practiced calorie restriction, made cryonics arrangements, trained as an Alcor cryonics transport technician – and who was murdered in North Africa as a result of exposing himself to excessive risk by his desire to 'confront death'. Concern for immediate safety should be the highest priority.

Another important problem with immortalism is that it is an affront to religion. It is not an affront to religion to speak of an extended life span of 100 years or even 1,000 years. After all, Methuselah reportedly lived 969 years. To say that a 1,000 year life span is an affront to God would be to insult God (if such exists). What is 1,000 years – or even one million years – to Eternity?

If human beings were free of disease and senescence the only causes of death would be accident, suicide and homicide. Under such conditions it is estimated that from a population of one billion, a 12-year-old would have a median life span of 1,200 years and a maximum life span of 25,000 years (i.e., one-in-a-billion would live the maximum 25,000 years). Thus, I can say that my goal is to live to one thousand. That does not mean I wouldn't like to live longer, but I will remain focused on my goal. It also means that both cryonics and anti-aging science are simply extensions of medicine, rather than a challenge to religion (cryonics patients have not died, they have deanimated). In practical terms, someone who claims to be trying to extend his or her life span may be less likely to be snuffed-out by a fundamentalist physician than a would-be immortalist who is seen as a blasphemist.

The more cryonicists can present themselves as life-extensionists rather than immortalists, the better the chance cryonics has of being accepted (or, at least, tolerated) by medicine, by religion and by society. The more cryonics is accepted, the better is the chance that cryonics and cryonicists can survive.

Some question that immortality may not be achievable because of such things as extinction of the sun, heat death of the universe, and proton decay. I find it difficult to even take a theoretical interest in these questions. The most immediate issues for our survival are staying alive as long as possible, the elimination of aging and disease and ensuring that cryonics can work. If these problems can be solved, we will have hundreds or thousands of years to think about other threats to our existence. If they cannot, other problems are irrelevant. If I am alive in a youthful condition 200 years from now, then the most awesome problems of mortality will have been solved – and the chances of finding ways to ensure survival for another 800 years will be trivial in comparison.

There is little to be gained by worrying about circumstances beyond 1,000 years. We cannot now comprehend the conditions of life and survival a thousand years in the future anyway, so it is a waste of effort to try. The most immediate survival goals are to either live long enough to benefit from tangible reversals of aging through technology or to see reversible suspended animation of the brain. That could happen in anything from 10 to 50 years.

For those who survive the next 50 years, during which the elimination of aging is bound to occur (in my opinion), the next challenge will be to survive death-by-accident and to learn to live safely. Close behind that danger will be death-by-murder, because the progress of science will always include the power for people to annihilate other people by increasingly sophisticated means. Following that problem will be self-annihilation through transformation. As people augment themselves with smart drugs, biological add-ons, computational and communications hardware, migration to other platforms, etc, they may easily lose their 'self' in the process.

Although it is true that the longer we live, the more adept we become at surviving, it is also true that we only need to be a victim of murder or a fatal accident once to be obliterated forever. However small we can make the probability, with enough time a fatal event is inevitable.

Is there anything to be gained by attempting now to take on the problems of survival beyond a millennium? Don't we have enough problems to cope with, without presenting ourselves as enemies of religion? Let's concentrate of extending our lives long enough to take the next step – or there might not be a next step to take. Let's be life extensionists – trying to survive the next thousand years – rather than 'immortalists'.

An Introduction to Immortalist Morality

Marc Geddes

The desire for immortality is one of the deepest, most enduring dreams of humanity. But is it a noble dream? Advanced technologies such as biotech, nanotech and infotech appear to hold great promise for extending human life spans and restoring youth at some point in the not-too-distant future. But even assuming that radical life extension is possible, some people find the idea disturbing. There are philosophers who argue that the quest for immortality is morally wrong, that we should accept aging and death as a necessary part of life. In this essay an argument for the opposite conclusion is presented. It shall be argued that not only is the quest for immortality morally good, it is in fact the very foundation of morality!

Moral Theory

Any theory of morality has to begin somewhere. We shall here begin with something known as 'moral intuitionism'. Moral intuitionism is the idea that some moral precepts are understood through direct conscious awareness, rather than through logical arguments. Appealing to the intuition of readers then, the proposed starting precept is very straightforward: "life is better than death."

Can it be logically proved that life is better than death? The question is debatable, but such proof is not necessary, provided that virtually all readers can agree that this is a good starting premise. One does not have to agree that in all circumstances life is better than death. Sometimes death may indeed be preferable. All that is being proposed is that in general, life is better than death. Most sane people could probably intuitively agree with the claim. In fact, the preference for life appears to be a universal throughout human culture. It is near universal for humans to celebrate birth and lament death.

Let us now apply the idea that life is generally better than death to the ethical question of human life extension. Suppose that at some point in the future science finds a way to eradicate aging and disease, so that – barring accidents or violence – a person could live on indefinitely. Let us further suppose that science can not only lengthen life, but also fully reverse any disabilities and symptoms of old age, so that everyone can enjoy the vigor of a healthy 20-year old. Put aside the question of whether or not such a thing is possible for the moment. The question we are asking here is whether or not eternal youth would be ethical. How long would you choose to live, if you had the choice of living in good health for as long as you desired?

A variety of possible objections to the offer of eternal youth present themselves. The objections can be divided into two different categories: practical and philosophical. Practical worries might include: the population problem, the problem of scarce resources and environmental pollution, eternal youth that is only available to the wealthy, the accumulation of too much wealth and power by an elite group of immortals.

We shall not here examine the myriad of practical problems that radical life extension might cause. It shall simply be noted that the historical record suggests that almost any scientific or technological advance causes new practical problems. (As an example take the Internet.) In the case of the question of radical human life extension, it is certainly reasonable to assume that should such a thing come about, it will cause problems. But all of these problems might be solvable. We need to examine the philosophical reasons for wanting life extension. If we find that there are strong ethical reasons for life extension, this means we can be more confident that life extension would be on balance a good, regardless of the problems that might arise from it.

While most people might accept that generally life is better than death, it has to be considered whether this is only true for a finite length of time. Perhaps life is better than death for a while, but then the postulate ceases to be true. Is there a time limit to the claim that life is better than death? It is hard to see why this should be so. If there is a time limit, where does it lie? If you think that living to 100 in good health is better than living to 50, why is 200 years of great health not better than 100? Why stop at 200? Why not 500? If you would be happy to live to 500, why not a million years, even forever?

A philosophical objection to life extension is the worry that the longer we lived, the less we would value our time. After all, a basic economic principle is that the value of a resource tends to increase the scarcer it is. Would we somehow value each moment less if we lived longer? Another worry that people may have is that a desire for life extension is somehow selfish. Perhaps budding immortals would become really self-centered and narcissistic?

It shall now be argued that both of these philosophical objections are without merit. It will be shown that not only does striving for a longer life increase the value of each moment, but it also increases the motivation for moral behavior.

VALUING INFINITE LIFETIME

The first stage of the argument follows from the fact that the continuation of our lives requires effort, both individually and on the social level. The continuation of human life is not guaranteed. There are basic survival requirements. Humans need air, water, food and shelter at an absolute minimum. We have to take actions on an on-going basis to ensure survival. Staying alive takes work! At any moment sentient beings have choices. Some of the choices we make will harm our chances of survival. Other choices will enhance our survival prospects. Since life is better than death, it follows that the choices that harm survival prospects are bad, and the choices that enhance survival prospects are good.

It is clear, however, that simply focusing on our own short-term survival hardly leads to other ethical behavior. For instance, we could steal someone's wallet. If there was a lot of money in it, it might enhance our own short-term survival prospects greatly, but few people would regard this as moral behavior.

But why should our goal simply be short-term survival? For the example of stealing someone's wallet, such behavior may help the thief in the short-term, but could it be that in the long run such behavior actually reduces survival prospects? Imagine if everyone lived in a barbaric way, trying to take advantage of everyone else. In the distant past social life was closer to this. Small tribes spent their time fighting with other tribes - rape and pillage were the preferred modus operandi. This pre-civilized state has come to be known as 'Hobbesian' after the political philosopher Thomas Hobbes.

What Hobbes pointed out was that it would actually be to everyone's long-term advantage to accept some limitations on their behavior. The idea was that people could always hurt each other if they were determined enough. If A hurt B, then B's angry friends could strike back. Be nasty to people and they are more likely to be nasty back. Tit for Tat. From Hobbes and Locke came the idea of a 'Social Contract'. [1] A contract is a formal or informal agreement whereby parties agree to recognize various obligations to one another. In the case of the 'social contract' the idea is that everyone living in a society has implicitly agreed to play by a set of rules because, over the long run, playing nice makes everyone better off than they would be in the anarchistic 'Hobbesian' world. These ideas form the basis of a political philosophy known as 'Contractualism'.

Rational people understand that actions have consequences. A life of crime may help a person in the short term, but in the long run it may get you killed or imprisoned. That we recognize it is in our own long-term interests to respect others leads to moral behavior. When we respect the rights of other people, they are more likely to co-operate with us, to mutual benefit. Of course, for this to work people have to learn to defer short-term gain in favor of long-term benefit. The critical point is a person's awareness that they have a future. People are more likely to be moral when they understand they will have to face the consequences of their actions in the future. It follows that the further into the future one plans for, the more moral one's behavior should become. People that live a short time do not have to experience the future consequences of all their actions. Longer lives should reduce the tension between the individual and society.

The extent to which moral behavior stems from the ability to plan for the future has perhaps not been properly recognized as of yet. Evolutionary biologists have tried to understand moral behavior such as altruism in terms of the possible survival advantages it would bring. But moral behavior cannot be fully explained in this way – if we only consider the short-term.

In a recent Nature paper [2], a moral dilemma was described. Two subjects were asked to share a pot of money, say $100, in which one person decided the amount that each of them got and the other person either had to accept it or neither of them would get any money at all. They only play once. If A decides that he should get $95, and offers B only $5, it might seem logical for B to accept. After all, B gets $5 if he accepts, and nothing if he does not. But when the game is played in real life, people refuse to accept splits that are too unequal, foregoing personal gain in order to punish the other guy.

Moral behavior is only of advantage when the game is played many times (an 'iterated game'). Over many games it is logical for the person who is offered an unfair split to decline, and for the person splitting the money to do so fairly. This is because it is known that the optimum strategy for interactions between two parties over the long run is simple 'Tit for Tat'. The success of the 'Tit for Tat' strategy was discovered in a worldwide computer competition in 1981. The competition was looking for a solution to a moral dilemma known as 'The Prisoner's Dilemma.'

There is a lesson there. In the real world kindness to strangers is only really to one's advantage over the long run. In fact, morality would only be perfectly logical if we lived forever. People have to stick around long enough to reap all of the consequences of their actions. When humans act morally they are in a sense acting as if they are immortal! We can conjecture that human morality is in part explained by the uniquely human sense of time. Only sentient beings capable of rational thought can plan far into the future, and only they can understand that the world will continue to carry on without them should they die. Humans are motivated to act morally because, in their imaginations, they can consider what people would think of them if they were alive at any time into the future – be it 5 minutes from now or 5 centuries.

Let us also consider the question of the value of the moments in a person's life. Time is no ordinary commodity! A person with more time can plan further into the future. They have more choices available to them in the present because some of the things a person could do in the present would only pay off over the longer-term. A person with more choices has, by definition, more freedom, and has an increased range of goals to choose from. Thus the longer a person has to live, the greater the potential value of each moment.

We have presented strong reasons for believing that life is generally better than death no matter how long an individual might live. Firstly, the potential value of each moment is increased the longer a person lives. Secondly, the longer a person expects to live, the greater the motivation for moral behavior. This would appear to clinch the ethical justification for life extension: Life extension is morally good. Since the arguments apply over any length of time, the longer an individual can potentially live, the greater the good. Thus it is actually an ethical imperative that we strive for immortality! Since a truly immortal person would live an infinite time, it seems that immortality is in a sense an infinite good. It would be a reasonable conjecture then, that the quest for immortality is the ultimate moral imperative. Let us call this idea 'immortalist morality'. The idea is that we base the whole of ethics on 'affirmation of life'. When sentient beings make a life-affirming choice, this is designated as morally good. When sentient beings make a choice that degrades life, this is designated as morally bad. Why not make immortalist morality the entire foundation of our value systems?

An important point to note here is that morality and legality are two separate spheres. There is a danger with any suggested ethical system that some people will want to impose these morals on everyone else. But the problem is here is one of tolerance. It is not being suggested that laws be passed on the basis on 'immortalist morality', although it could serve as guide in formulating public policy for some issues.

A number of philosophers have tried to construct ethical systems by taking affirmation of life as the foundation for morality. Ayn Rand based her Objectivist theory of ethics on the idea that one's individual life is one's ultimate value. The German humanitarian and theologian Dr. Albert Schweitzer wrote:

> Affirmation of life is the spiritual act by which man ceases to live unreflectively and begins to devote himself to his life with reverence in order to raise it to its true value. To affirm life is to deepen, to make more inward, and to exalt the will to live. At the same time the man who has become a thinking being feels a compulsion to give to every will-to-live the same reverence for life that he gives to his own. He experiences that other life in his own. He accepts as being good: to preserve life, to promote life, to raise to its highest value life which is capable of development; and as being evil: to destroy life, to injure life, to repress life which is capable of development. This is the absolute, fundamental principle of the moral, and it is a necessity of thought. [3]

A major objection to the idea that ethics is derived from the survival goal, is that there are many things that we value in life beyond mere survival. After our physical needs are taken care of, we still have physical desires that may conflict with our survival. And quite apart from physical desires, we have emotional and intellectual goals. Is it not better to regard survival as just one value out of many, and ethics as a weighing up of multiple preferences?

It is important to understand that many of our desires are actually by-products of evolution. Evolutionary psychology studies the ways in which human actions may be driven by biological urgings that evolved because they provided an evolutionary advantage. For instance, our emotions and our thoughts indirectly aid our survival prospects. Emotions enable us to empathize with others, helping us to better co-operate with people. Social skills provide a clear survival advantage. Rational thinking is a survival tool, because we can use abstract thinking to understand and predict how the world works. 'Life affirming' things can even include art and philosophical systems of thoughts. We have come to value these for their own sake, but they arose by chance and remained because they provided an evolutionary advantage for humans. Of course it was genes that evolution was selecting for, not the extension of individual lives. But survival needs did tend to correlate with reproductive fitness. So making the survival goal primary does not necessarily conflict with the many other things we come to value. Immortalist morality conjectures that everything worthy of value stems from the quest for immortality.

STAGNATION

The most common philosophical objection to radical life extension is that really long life would simply get too boring. Perhaps we will simply run out of interesting things to do? Would we end up in a static world where there is nothing new under the sun? But the opposite is argued here. If everything worthy of value comes from the quest for immortality, then a long life should actually be more interesting and meaningful. How could this be?

The first point to note is that once technology becomes advanced enough to radically extend human lifespan, it is likely that technology will also be advanced enough to radically alter the minds and bodies of those who desire it. For instance we could imagine 'brain refresher drugs' which prevented brains from becoming too inflexible. The people living in the far future might be able to alter their bodies and personalities as easily as the people of today change their clothes. The fact that some people living today get tired of life is more likely to be a practical, biological problem than a philosophical one.

IDENTITY

Another philosophical concern emerges. An individual might worry that if he lived long enough he would cease to be 'himself' and become someone else. After all, exactly what is the 'self'? Astronomer Martin Rees recently expressed this worry:

> I'm reconciled to extinction – losing all consciousness as well as rotting away physically. Indeed, I think we should welcome the transience of our lives. Individual immortality would be deleterious for life's further development unless we could transform ourselves, mentally and physically, into something so different from our present state that the transformed entities wouldn't really still be 'us'. If technology allowed me to transcend these limitations, I would only be the same person in the sense that I would retain some memories of early life. But even over present life spans it is not clear how much continuity of personality is really preserved. Each of us is a 'bundle of sensations' somehow woven together as a continuous thread or 'world line'. [4]

The idea that we are just a 'bundle of sensations' dates back to philosopher David Hume, but other philosophers such as extropian transhumanist Max More would disagree, regarding the concept of 'self' as perfectly meaningful. [5] The philosophy and science of the mind is currently not advanced enough to provide an answer to the question of the self. So long as a living being retains memories of his past, there is a connection between its past and present selves which may be sufficient for it to retain the same sense of 'self'. In order to remember our past, our current self has to be 'backwards compatible' with all our past selves.

A related worry is that an extremely long-lived person would somehow cease to be human. Yet human nature itself is not fixed. Human beings have constantly re-invented themselves through cultural change and new technology argues Transhumanist philosopher Nick Bostrom. [6] Even if it would be true that someone who lived hundreds of years started to change into a different entity, why should this be feared? After all, a man at 20 years is rather different to a man at 5 years, just as the man at 60 years is rather different to the man at 20. But the potential for change is precisely what makes life exciting and creates the opportunity for something better to come along. And consider the alternative: death. Did we not agree that life is generally better than death? Better to evolve than die!

Religion

Some people may object to the quest for immortality on religious grounds. It is argued that extremely long life is somehow un- natural, that it is 'against God's plan'. Yet some of the strongest allies of the quest for immortality may come from those of Jewish faith. In Judaism the primary metaphor for God is that "God is Life". Judaism may be the religion most compatible with 'immortalist morality'. At a 1999 conference on life extension, Rabbi Neil Gillman had this to say:

> There is nothing redemptive about death. Death is incoherent. Death is absurd.

The rabbi was asked if Jewish tradition would endorse prolonging human life for twenty years. "Yes", answered the rabbi. Forty years? "Yes." One hundred years? "Yes." He regarded the indefinite prolongation of life as a moral good. [7]

MOTIVATION

What would motivate very long-lived people to continue to strive to create new things and explore new realms? The basic moral premise we have been talking about: the desire to see life survive. This is an ongoing process: it is a journey not a destination. If immortality was something that we reached at some point, then it could no longer serve as the foundation of ethics. But no matter how far advanced our science and technology becomes, it is unlikely that the continuation of our own life can ever be guaranteed. It may be theoretically possible to live forever, but this would likely involve the continual solving of new problems and overcoming of new challenges. We could think of guaranteed infinite life span as a sort of mathematical limit, which we can get closer and closer to, but never quite reach. Each new scientific advance would lower the risk of dying, but the quest for immortality would continue forever.

Would not people who adopted an immortalist morality become very risk adverse? The answer is no, because people have to take some risks in order to continue to survive. There is no paradox between aiming to live a long time and taking some risks in the short-term. Short-term risks have to be taken to secure long-term gain. For example, a person would not get up out of bed in the morning if they were trying to maximize their short-term survival chances. However, a little longer-term thinking can determine that the rational action is to take the risk of getting out into the world and accomplishing your goals. That is the only way to make ends meet and survive over the long run. It is important to emphasize that staying alive requires constant effort, and immortality is a journey not a destination.

SELFISHNESS

There is another point to consider here. Immortality as a fundamental moral imperative need not be interpreted in a purely egotist sense. That is, it is not being argued that the survival of our own individual life is central at all times. We can also allow that the lives of others may take precedence in some circumstances. It is perfectly consistent with immortalist morality to take some risks in order to help the survival chances of others. Throughout history, explorers and soldiers put their lives on the line for the good of others. In some circumstances it may be life-affirming to practice altruism, and sacrifice our own life so that others may live. So clearly, immortalist morality is not totally selfish.

This also answers the objection that the chance of radical life extension in our lifetime is too low to make it a worthy goal. Some may say that it is not worth wasting time on such an unlikely goal when there are many problems to solve in the here and now. But our own personal chance of success has no bearing on whether or not life extension is a worthy goal. Even if we die long before scientific advances find a cure for old age, the goal is still worthy in so far as we are helping others to reach it.

It is interesting to point out that the suggested foundation for morality can be empirically falsified. Perhaps we cannot prove that the quest for immortality should be the ultimate value, but we could disprove it. If science ever determined that it is impossible for life in the universe to last forever, the quest for immortality is impossible and cannot be justified. So, does the scientific evidence rule out the idea that life could, in principle, survive forever? Some have thought so. One argument that life cannot survive forever comes from a law of physics known as the second law of thermodynamics. This says that the entropy of an isolated system (a system which exchanges no matter or energy with its environment) must always increase. Entropy is a measure of how disordered the system is. For instance science writer Adrian Berry once wrote: "Preserving a living body forever would violate the second law of thermodynamics." [8]

In fact the second law of thermodynamics does not imply that a living thing has to decay. Living things are not isolated systems. They constantly exchange matter and energy with the environment. For instance the human body excretes waste and takes in air, food and water. So long as a living thing continues to take in new energy, there is no reason why it has to decay. The biosphere of planet Earth as a whole is exchanging energy with the wider solar system. What about the universe as a whole, however? The universe appears to be an isolated system in which entropy has to increase. Will all sources of useable energy one day run out? Will everything decay? The great philosopher Bertrand Russell certainly thought so. He wrote these depressing words:

> "That man is the product of causes that had no prevision of the end they were achieving; that his origin, his growth, his hopes and fears, his loves and his beliefs, are but the outcome of accidental collocations of atoms;

that no fire, no heroism, no intensity of thought and feeling, can preserve individual life beyond the grave; that all the labors of the ages, all the devotion, all the inspiration, all the noonday brightness of human genius, are destined to extinction in the vast death of the solar system, and that the whole temple of Man's achievement must inevitably be buried beneath the debris of a universe in ruins- all these things, if not quite beyond dispute, are yet so nearly certain that no philosophy which rejects them can hope to stand. Only within the scaffolding of these truths, only on the firm foundation of unyielding despair, can the soul's habitation henceforth be safely built." [9]

However, to a scientist, it is not known how the universe will end, if at all. It seems that the universe might go on expanding forever, but the nature of dark energy is not well enough understand to be very sure. It is important to understand that even if the average entropy density of the universe has to inevitably increase, this does not mean that it ever has to reach a maximum and stop dead. Even if the universe comes to an end, it may still be possible for life to survive forever. In 1979, English physicist Freeman Dyson published a paper [10] in which he argued that even in a universe with finite energy an intelligent being could still think an infinite number of thoughts. He considered the case where the universe kept expanding, but started to 'die' as useable energy ran out. He found that as the universe grew colder and colder advanced beings could still live forever by thinking thoughts at a slower and slower rate. Physicist Frank Tipler considered the opposite scenario - the universe one day stops expanding and starts to collapse under the force of gravity, coming to an end at a 'big crunch'. His idea was that as the universe grew hotter and hotter, intelligent beings could still live forever by thinking thoughts at a faster and faster rate, but only if they developed technologies powerful enough to modify the large scale structure of the universe. This is known as the 'Omega Point' theory. [11]

It emerges, that the empirical data are insufficient to determine whether life in the universe has to end. So there is no scientific basis for Russell's pessimism and we can advance as a reasonable conjecture the claim that life can last forever.

It is interesting to note however, that life on Earth will probably have to expand into space in order to continue to survive. If life on Earth has to one day expand into space to stay alive, and continue expanding, then it seems unlikely that life would ever become boring or free of risk, no matter how long one lived. There will always be exciting new challenges to face and it is precisely the quest for immortality that will drive humanity to face them! This is all the more reason for believing that the quest for immortality should indeed be the ultimate moral imperative.

It has been argued that immortality is possible, but only if rational beings make continual efforts to stay alive. It is not being argued that immortality can ever be guaranteed. If science ever found a way to guarantee immortality, immortality would cease to be a goal and could not provide the basis for morality.

Tipler's idea that life will one day have to spread across all space and develop technology powerful enough to change the structure of the universe is intriguing because it suggests that the very fate of the universe is tied to the efforts of living things to stay alive. If so, immortality could be said to be the very 'telos' (end purpose) of the universe.

References

1) Hobbes, Thomas; 1642, "De Cive" [On the Citizen] // Locke, John; 1690, "Two Treatises of Government"

2) Fehr, Ernast & Fishbacher, Urs; "The Nature of Human altruism" in: *Nature* (23 Oct 2003, 425), pg. 785–791

3) Schweitzer, Albert; *Out of my life and thought,* (1953) John Hopkins University Press

4) Rees, Martin; *ImmInst Interview* (2003); http://imminst. org/forum/index.php?act=ST&f=67&t=2699&hl=&s=

5) More, Max; *The Diachronic Self* (1995); http://www. maxmore.com/disscont.htm

6) Bostrom, Nick; *In Defense of Posthuman Dignity* (2003); http://www.nickbostrom.com/ethics/dignity.html

7) From the *Extended Life/Eternal Life* conference at the University of Pennsylvania; (6. March 1999)

8) Berry, Adrian; *The Next 500 Years* (1995) Headline Book Publishing

9) Russell, Bertrand; "Appendix B: The Doctrine of Types," in: Russell, Bertrand; *Principles of Mathematics,* (1903); Cambridge University Press; pg.523–528

10) Dyson, Freeman; "Time Without End: Physics and Biology in an Open Universe" in: *Reviews of Modern Physics* (July 1979, Vol. 51, No. 3)

11) Tipler, Frank; "Cosmological Limits on Computation" in: *International Journal of Theoretical Physics* (1986, 25), pg. 617–661

Should We Fear Death? Epicurean and Modern Arguments

Russell Blackford, Ph.D.

Most of us fear death, to a greater or lesser extent, though some philosophers believe that we would do well to accept it and to fear any prospect of immortality. Bernard Williams, in particular, has argued that we would eventually suffer unbearable boredom, and come to welcome death, if we had the ability to live for hundreds of years (p89–98). [1] Though we might die earlier than we would like, he suggests, the fact that we all die is actually a good thing. Many others have argued, ever since antiquity, that death is at least not something to be feared.

In this essay, I argue that it is rational to be attached to life and live as long as we can, though not to fear death with the intensity, or nagging anxiety, that human beings often do. Furthermore, our reasons for being attached to life are also reasons why we should want to live indefinitely.

In the ancient world, the first philosophical attacks were made on the rationality of fearing death, based on the assumption that there is no afterlife and that death extinguishes all sensation, thought, and awareness. Separate issues arise, if we have religious grounds to believe that there is an afterlife of eternal bliss or punishment. I will set those aside and consider the fear of death and our attachment to life purely from a secular philosophical viewpoint. (I will also set aside whether it is rational to fear the process of dying, as opposed to death itself – though there is no doubt that the process is usually painful and nasty.) The locus classicus of the debate is the work of the Hellenistic philosopher Epicurus and his followers, who viewed death in a strikingly modern way, as the end of all sensation or awareness. On that assumption, we cannot rationally fear it as a great unknown, or as a prequel to divine judgment and possible punishment. Is there any other rational justification to fear death, or consider it a bad thing? What is so bad about death?

ARGUING LIKE AN EPICUREAN

In his *Letter to Menoeceus*, Epicurus argues "all good and evil lie in sensation" [2; pg.149]. Since death is the extinction of sensation, it is "nothing to us", something that is neither good nor evil. This can be formulated as what I will call The Basic Epicurean Argument:

> The Basic Epicurean Argument:
>
> P1. Nothing is a misfortune unless it includes or causes unpleasant sensations.
>
> P2. Death does not include or cause unpleasant sensations.
>
> C. Death is not a misfortune.

This argument is logically valid. More interestingly, it is amenable to significant modification should counterexamples be offered to challenge P1. This premise can be altered in numerous ways without affecting the validity of the argument, so long as appropriate alterations are also made to P2.

By way of explanation, consider the following counterexample to P1: a close friend who loves me ceases to do so (for whatever reason). I might never be adversely affected by this, in the sense of suffering unpleasant sensations, e.g., my friend might maintain a pretense of love, and I might, as events turn out, never even become aware of the change in her feelings. Yet, the loss of a friend's love is usually considered to be a misfortune. [3; pg. 4–6]

An Epicurean could respond to such counterexamples by making suitable modifications to both premises of The Basic Epicurean Argument. Thus, the Epicurean might point out that, if a friend has ceased to love me, she might thereafter have some propensity to act in ways that I will find unpleasant. The Epicurean could modify P1 by including among the classes of misfortunes those events that, at the time they happen, make us more vulnerable to unpleasant sensations. She could then plausibly modify P2 to state that death is not such an event – after all, I will have no unpleasant sensations, or sensations of any other kind, once I am dead. Since both premises have been modified appropriately, the argument remains valid.

Not all possible counterexamples can be accommodated in this way. For example, will it not be a misfortune for me if my reputation is defamed in some way after I die (possibly as a consequence of my death, since I will no longer be able to defend myself). An Epicurean could respond to this kind of example with a second strategy. She might suggest that the new example is not a misfortune. Rather, someone who worries about such things is in the grip of a kind of pride that is irrational because it is not conducive to living the happiest kind of life.

This brings us to the nub of the matter. A full Epicurean argument against the rationality of fearing death would have to include a specific account of the good life.

The Epicureans had such an account and it actually has some attraction. Specifically, they claimed that the good life consists in living with freedom from pain (aponia) and freedom from anxiety (ataraxia). They believed that achieving aponia and especially ataraxia (which we might translate as "tranquility") for as long as we are alive and conscious, is the highest level of human happiness, and that many of our strivings and concerns actually frustrate our happiness. Such propositions can be found in Epicurean documents such as the *Letter to Menoeceus* and, particularly, the *Key Doctrines* (149–150). [2] The Roman poet Lucretius, the greatest of Epicurus' followers, argues in detail in his masterful *De Rerum Natura* that our lives are blighted whenever we strive after goals that are inconsistent with ataraxia (p 151–153). [4]

If we could accept these more general Epicurean views, we might well limit the classes of genuine misfortune and we might conclude that a combination of the strategies that I explained above could protect the spirit of The Basic Epicurean Argument against any possible counterexample to P1. This is because, according to the Epicurean conception of happiness, any genuine misfortune must be something that can interfere with our tranquility while we are alive. Death itself does not do so – though fearing it does – so death itself is not a misfortune. To an Epicurean, then, death is not a bad thing.

On the other hand, this analysis suggests that The Basic Epicurean Argument can be successfully defended only if we accept general Epicurean views about the nature of happiness. These would require us to jettison many of our commonsense ideas; such as that it is a misfortune to be despised by others (even if we are blissfully ignorant).

It appears to me that there is more truth to the Epicurean view of happiness than is usually acknowledged. There does actually seem to be a limit to our happiness, and we do seem to get close to that limit if we find ourselves living healthy lives without pain or anxiety. Much of our striving for more than that seems futile, or even counterproductive. Still, I will argue that there is more to a good life than the Epicureans articulated in their philosophy. That 'something more' is what makes it undesirable to die.

DEATH AS A DEPRIVATION OF ADDITIONAL LIFE

We could reject the Epicurean account very quickly if we insisted that being alive, as such, is a good thing, and saw death as a misfortune simply because it deprives us of additional life. That, however, creates more problems, since the concept of deprivation is not straightforward. It appears to include the idea of being denied something that it was possible to have. However, in what sense is it possible for a person to have a longer life than (speaking without tense for the moment) she actually does have? That question raises intractable issues about determinism, fate, and free will, issues that it seems better to avoid if we are to make any progress.

To avoid them, I want to focus more closely on the concept of fearing death. There is a relationship between fearing some kind of future event and acting to avoid or resist it. For example, we try to avoid disease by inoculations, healthy diet, good hygiene, etc. Similarly, we may avoid violence by fleeing it, or we may use force to resist. Even if the Universe is deterministic, our own actions to avoid or resist such things must surely form part of the chain of deterministic causes. If so, actions such as fleeing or opposing violence are rational.

Thus, whether or not determinism (or fate) prevails in the Universe, there are many actions that we can take to avoid or resist events that would otherwise be the death of us. If additional life is a good thing, it is rational to take such actions.

On that assumption, it seems rational to act as if we feared death. If we had the choice, we would even be prudent to 'program' ourselves to have some degree of subjective fear of death, in order to motivate ourselves to identify and undertake actions that will help us live longer. It does appear to me that some fear of death is a good thing. Later, however, I will ask how much fear of death we ought to have.

None of this reasoning is sound unless being deprived of additional life is, indeed, a misfortune and there is a ready reply to that claim. Once again, our friends, the Epicureans, made it in antiquity.

Do we Want Additional Life?

In *De Rerum Natura*, Lucretius argues that it is not bad to be deprived of additional life, since no one thinks it was bad not being alive, prior to the time we were born. We are not, he suggests, upset about not having been alive at the time of the Carthaginian wars (p151). [4] If missing out on additional life after our deaths is a misfortune, what about all that life we missed out on before we were born? Of course, it is not rational to fear a misfortune that is already in the past. It is only rational to fear future events. However, an Epicurean could add, past misfortunes are surely matters for regret. Yet we do not even regret not being around in the days when the Romans fought the Carthaginians.

Some attempts to answer this line of argument raise metaphysical issues about the nature of the self. Unfortunately, those issues—much like those to do with the prevalence in the Universe of determinism or fate — are too complex to resolve here. For example, Nagel argues that it is not possible for us to have any sensations very long before we were born – certainly not before we were even conceived (p8). [3] This assumes that each of us is the biologically organized matter continuous with a particular zygote, created at a particular point in time. Kaufman takes a different approach. He argues that even if the self, as a "metaphysical entity" (p309), [5] is defined in this way, the real problem for Lucretius is that events in my past affect my current personality. If I had been alive at the time of the Carthaginian wars, I would not now be as I now am, thinking the thoughts that I do and having the wishes that I have! That being so, Kaufman says, I cannot coherently regret not having been alive earlier, for I would be wishing away my own personality (307–311). [5]

In my view, these arguments are not at all decisive against the position of Lucretius. The short answer to Kaufman is that his approach proves too much. Many events whose occurrence we later regret influence the development of our personalities. Indeed, Kaufman's argument appears to rule out the appropriateness of regret in those cases where it actually seems to be most appropriate, such as when someone's personality has changed deeply over the years as a result of dwelling upon a crime, or some non-criminal but seriously hurtful act, that she once committed.

The argument put by Nagel is both more difficult and more convincing. In the end, however, it appears to fail. Assume for the sake of argument that 'I' am a biologically organized four-dimensional being whose earliest temporal stage is the zygote that was formed when one of my father's sperm cells fertilized one of my mother's ova. It seems that I could not simultaneously have an identity defined in such a way and have had experiences before I was even conceived.

At the same time, we can imagine logically possible scenarios in which the very same zygote could have come into existence at an earlier time than it did. Imagine that a certain person was born twenty years after a particular ovum and sperm cell were selected for in vitro fertilization. For all those years, prior to the IVF procedure being carried out, the two gametes were separated and cryonically preserved. The sperm cell was then used to fertilize the ovum, and, after further medical procedures (involving, let us say, a volunteer surrogate mother), the person's birth finally took place. Would she regret the 'lost' years when she might have been alive if the IVF procedure had not been delayed?

I doubt that regret would be either appropriate or actually experienced in the normal course (if this scenario can be thought of as in any sense normal). However, it is possible to imagine specific circumstances in which regret might be appropriate. What if the gametes were preserved near the end of a time of peace and prosperity, and the person concerned reached her adolescence just as this was ruined by a terrible war? Perhaps, in specific circumstances such as those, she would have cause for regret, but the regret would be that her adolescent experiences took place in a worse social environment than might have been, not that she missed out on an additional period of life.

Reflection on the circumstances in which we could rationally and consistently regret not being born earlier seems to me to strengthen the Epicurean arguments, at least to this extent: What is regrettable is not the mere fact of not being alive for a period of time. Still, I think that Lucretius can be answered.

Projects, Relationships, Commitments

A better reply to Lucretrius than that given by Nagel or Kaufman is the following, based broadly upon the views of Williams (p85–87) and (particularly) Martha Nussbaum (318–320), [1;6] even though neither would actually favor human immortality. The gist of the reply is this: Once we are born and begin to become part of a society, we soon have good reasons for preferring to stay alive, reasons that are forward-looking, so there is no symmetry with our the past before we were conceived or born.

I might, for example, wish to complete a novel or a work of philosophy that I am writing. I might be obsessed with the fluctuating fortunes of a favorite athlete or sports team. I might be involved in an interesting and charming flirtation, or perhaps an ever-deepening love affair, and there might be various people who are dependent on me, emotionally or financially.

In short, I might have a multitude of projects (some deeply serious, some less so), relationships (likewise), commitments, and interests that I can imagine extending and changing into the indefinite future. All of these are attachments to life, and almost everybody forms them. Even Epicurus died with a request to a friend to "take care of the children of Metrodorus!" (p151) [2] He was not entirely indifferent to what would happen after his death. Indeed, none of us could be indifferent to our own prospective deaths, what might follow them, and what they would prevent, while simultaneously retaining such attachments.

Nor would we be better off without such attachments to life. Lucretius is doubtless correct that foolish obsessions can distort our lives and lead to unhappiness (p151–153). [4] Yet, our forward-looking projects, relationships and commitments are an important part of what is valuable in our experience.

To the extent that the Epicurean conception of a happy life asks us to give them up, or to deprecate their importance, that conception is impoverished and should be rejected. Although there is some wisdom in the Epicurean worldview, it is incomplete at best. It cannot be used to save what I called The Basic Epicurean Argument against fearing death, because there is more to a good life than freedom from pain and anxiety – aponia and ataraxia. A good life includes rich kinds of experience and connection that are, by their natures, forward-looking.

None of this is to say that there is no force at all in the Epicurean arguments. In his *Key Doctrines*, Epicurus argues that "the flesh" (p151) wants infinite duration of life, while the intellect knows better. [2] In fact, I left open the question, earlier in this essay, of how much we should fear death, even if we have reasons to avoid and resist it. Perhaps "the flesh" has its own reasons to fear death, reasons that make us fear it more than is good for our happiness.

Though it is beyond the scope of this essay to pursue the issue in detail, the equivalent of "the flesh" in contemporary thought may be our unconscious genetic predispositions, some of which may have increased our evolutionary ancestors' inclusive fitness, but do not increase our happiness as individuals. It is rational to have projects, relationships, commitments, and interests that attach us to life. However, the degree to which we actually fear death – the sense of nagging anxiety or even panic that the thought of death sometimes causes – might not be something we would choose if we could reach into ourselves and rewrite our own genetic code, in order to harmonize our personalities with our considered ideas of what constitutes a happy life.

If my argument so far is correct, it would be rational to fear death less than we do, but it is also rational to want to remain alive, at least for as long as we have projects, relationships, etc., that attach us to life. Moreover, it is also rational to feel frustrated when we experience, or imagine, the decline in our mental and physical powers that will make us less and less able, as we age, to carry out our projects and commence new ones. Indeed, our knowledge that aging and death await us restricts what projects we can rationally commit ourselves to in the time available. If not for the specters of age and death, we could commit ourselves to projects that might take hundreds of years to see completion.

Williams' argument that we would experience terrible boredom if we could live forever (p89–98) [1] strikes me as rather unconvincing. As long as I have my full capacities, I can see no limit to my ability to immerse myself in new projects, in new and more relationships, in new interests. I suppose it might be different if the world did not change around me, so that a time came when nothing was new. But why should that ever be the case? Technology will advance, society will change, our understanding of the Universe will deepen, and we will find time to explore it.

I suppose one counterargument is that the longer our lives become the less we can have vivid and immediate memories of our entire lives. Very long-lived people might have some difficulty maintaining a psychological connection with their pasts, for, in a greatly extended life, memory may not be able to handle all that has been experienced. However, the extent of this problem is unclear, since we know so little about the neurophysiological workings of memory.

In any event, it is not obvious that the outcome would be terrible – or drastically different from everyday experience even now. I can remember little of my life before the age of five and I find that memories, even of critical experiences, become unexpectedly vaguer as I grow older. However, that does not mean that I fail to recognize how they have shaped me, nor has my past lost its interest to me.

If we live much longer lives than have ever been possible, our experience of our own identity may change, but that does not mean that our lives will be empty, discontinuous or without fascinating internal connections. Changed conceptions of personal identity might open up possibilities for growth, rather than being burdensome.

Perhaps if we could all live indefinitely, a time might eventually come for each of us when we believed ourselves fulfilled and no longer able to grow. Eventually, we might cease to feel attached to life in the ways that I have described, especially if we reached some kind of hard limit to our ability to adapt to a changing world, to maintain a strong sense of identity, and to remain creative. Personally, though, I would like to have the option of exploring the world and my own capacities until I found those limits. If I ever did come up against them, I could die at a time of my own choosing. If everything was right with my health, and my creative and intellectual powers, I can not imagine that any such time would arrive except many, many years after the completion of the eight or ten decades that I can currently hope for.

When we take a hard look at what is so bad about having to decline and die, it appears that we fear death too much. Yet, we should be careful not to rationalize the situation to the extent that we understate what is regrettable about aging and death. The reasons why death is a bad thing are simply the reasons why we are attached to life. Not only that, they are reasons that could attach us to it indefinitely, at least if we could retain the faculties that enable us to live life to the full.

This suggests that we should do what we can to turn down the volume of our fear of death, but we should not console ourselves with false reassurances about the supposed virtues of being mortal. Certainly, we should not adopt a stance of fashionable pessimism about the desirability of living beyond the rather pitiful limits that nature has allowed. Instead, we should

live to the full, pursue our projects, and enjoy our relationships and interests for as long as we can, with no equivocation
or apology. If we can extend the span of robust human life,
or even live indefinitely, that is a prospect to embrace with all
our optimism and energy.

References

1) Williams, Bernard; "The Makropulos Case: Reflections
on the Tedium of Immortality" in: *Williams' Problems of
the Self: Philosophical Papers 1956–1972* (1976, paperback
edition), Cambridge Univ. Press; pg.82–

2) Epicurus; 1987, extracts from *Letter to Menoeceus,
Vatican Sayings,* and *Key Doctrines* in: *The Hellenistic
Philosophers, Volume 1: Translations of the Principal Sources,
With Philosophical Commentary* (1987); Long and Sedley,
Cambridge Univ. Press, pg.149–

3) Nagel, Thomas; "Death" in: *Nagel's Mortal Questions*
(1979) Cambridge Univ. Press ; 1–

4) Lucretius; *De Rerum Natura* in: *The Hellenistic
Philosophers, Volume 1: Translations of the Principal Sources,
With Philosophical Commentary* (1987) Long and Sedley,
Cambridge Univ. Press, pg.151

5) Kaufman, Frederik; "Death and Deprivation; Or, Why
Lucretius' Symmetry Argument Fails" in: *Australasian
Journal of Philosophy* (1996, vol. 74); pg.305–

6) Nussbaum, Martha C; "Mortal Immortals: Lucretius
on Death and the Voice of Nature" in: *Philosophy and
Phenomenological Research* (1989. vol 50); pg.303–

CHAPTER III: RESOURCES

We conclude this introduction to the ongoing scientific conquest of death, with a challenge.

We invite you to
- investigate any of the topics raised, starting with the selected references in the bibliography.
- learn a bit more about the numerous authors and their work.
- visit us, to learn more and to share your views on the exciting project that is the scientific conquest of death.

Please note:
These resources, more information, and accompanying material are available at the website which has been set up for the special benefit of readers of this book:

www.ImmInst.org/book1

"Who wants to live forever?"

An interactive challenge to the reader

For those rooting for a breakthrough in life extension research, to question why it would be desirable to lead a longer and healthier life might seem banal. But a number of people cannot seem to conceive of any reason why anybody would want to live beyond the currently fashionable limit of about four score.

Some possible answers include:

- Watch your grandchildren and great-grandchildren grow up
- Find out what the future will be like
- Because art and creativity are inexhaustible
- Have more time to help others
- Why not?
- If you live, you can always change your mind about it later; death is irreversible.
- Watch Tibet beat Brazil in the football world cup final
- More time to figure out the meaning of life, if there is one
- Because it would suck to be in the very last generation to die of old age
- There are people who love you and who need you

- Have a chance to really grow up and find out what kind of wisdom and maturity might be attainable by a healthy 800-year-old
- Spend more time with friends and loved ones without a time bomb ticking quietly inside you all the while
- Learn the answer to some of the great mysteries: How does the mind work? Is there extraterrestrial life?
- Play, create, and make love, to explore exotic mental states
- Build and experience virtual realities
- Live happily ever after

To share and discuss your own thoughts visit
http://www.imminst.org/why

Inspirations by Nick Bostrom, PhD, Oxford University

BIBLIOGRAPHY

Why do we age?

Austad, S; *Why We Age: What Science Is Discovering about the Body's Journey Through Life*; (1997) ISBN:0471148032

Carey, J. R; *Longevity : the biology and demography of life span*. Princeton, (2002) ; Princeton University Press.

Finch, C.E; *Longevity, Senescence, and the Genome* (1990, second printing 1994); University of Chicago Press

Gavrilov, Leonid A; "Pieces of the Puzzle: Aging Research Today and Tomorrow" in: *Journal of Anti-Aging Medicine* (2002, Vol. 5); pg. 255–263

Hayflick, Leonard; *How and Why We Age* (1996 reprint edition); Ballantine; ASIN:0345401557

Kirkwood, Tom; *The Time of Our Lives: The Science of Human Aging* (1999); Oxford University Press ISBN:0195128249

West, Michael D; *The Immortal Cell: One Scientist's Quest to Solve the Mystery of Human Aging* (2003) Doubleday ISBN:0385509286

Anti- Aging Biomedicine: basic

Arnold, Nick & Benton, Tim; *How to live forever* (2000); Franklin Watts Inc. ISBN: 0531148181

Bova, Ben; *Immortality: How Science Is Extending Your Life Span and Changing the World* (2000); Quill; ISBN:0380793180

Benecke, M; *The dream of eternal life: biomedicine, aging, and immortality* (2000); Columbia University Press.

Bowie, Herb; *Why Die?: A Beginner's Guide to Living Forever* (1998); Power Surge Publishing; ISBN:1890457078

Gems, D & Pletcher, S & Partridge L; "Interpreting interactions between treatments that slow aging" in: *Aging Cell* (2002, Vol.1); pg. 1–9

Mc Gee, Glen; *The new Immortality: Science and Speculation about Extending Life Forever* (1995); Publishers Group West; ISBN:189316327X

Medina, John J; *"The Clock of Ages: Why We Age, How We Age, Winding Back the Clock"* (1997); Cambridge University Press; ISBN:0521594561

Olshansky, Jay S & Carnes, Bruce A; *The Quest for Immortality: Science at the Frontiers of Aging* (2003); W.W. Norton; ISBN:0393048365

Rubenstein, R & Benecke, M; *The dream of eternal life: biomedicine, aging, and immortality* (2002); Columbia University Press; ISBN: 0231116721

Shostak, S; *Becoming Immortal: Combining Cloning and Stem-Cell Therapy* (2002); State University of New York Press; ISBN:0791454029

Tandy, Charles; *Doctor Tandy's First Guide To Life Extension and Transhumanity* (2001); Universal Publishers; ISBN:1581126506

Tennant, Rich & Bortz, Walter M; *Living longer for dummies* (2001); Hungry Minds ISBN:0764553356

Various, (Sage Crossroads); *The Fight Over the Future: A Collection of SAGE Crossroads Debates That Examine the Implications of Aging-Related Research* (2004); iUniverse Inc. ISBN:059531631X

West, Michael D; *The Immortal Cell: One Scientist's Quest to Solve the Mystery of Human Aging* (2003); Doubleday; ISBN:0385509286

Wyke, Alexandra; *21St-Century Miracle Medicine: Robosurgery, Wonder Cures, and the Quest for Immortality* (1997); HarperCollins Publishers; ISBN:030645565X

Anti- aging biomedicine: advanced

Butler, RN & Fossel, M & Harman, SM & Heward, CB & Olshansky, SJ & Perls, T & Rothman, J & Rothman, SM & Warner, HR & West, MD & Wright, WE; "Is There an Antiaging Medicine?" in *Journals of Gerontology* (2002, Vol. 57); pg. B333-B338

Carnes, BA & Olshansky, SJ & Grahn, D; "Biological Evidence for Limits to the Duration of Life" in: *Biogerontology* (2003, Vol. 4); pg. 31–45

Freitas Robert A; *Nanomedicine, Vol. IIA: Biocompatibility* (2003); Landes Bioscience; ISBN:1570597006

International Association of Biomedical Gerontology. International Congress (9th : 2002 : Vancouver B.C.) & D. Harman (2002). *Increasing healthy life span: conventional measures and slowing the innate aging process*; New York Academy of Sciences

International Association of Biomedical Gerontology. (2001) & Park, SC; *Healthy aging for functional longevity: molecular and cellular interactions in senescence*; New York Academy of Sciences

Komender, J; "Stem cell research as a base for reconstructive medicine" in: *Annals of Transplantation* (2003, Vol. 8); pg. 5–8

Lo, KC & Chuang, WW & Lamb, DJ; "Stem cell research: the facts, the myths and the promises" in: *Journal of Urology* (2003, Vol. 170); pg. 2453–8

Toussaint, O; *Molecular and cellular gerontology*; New York Academy of Sciences.

Digitalisation

Kurzweil, R; *The Age of Spiritual Machines: When Computers Exceed Human Intelligence* (2000); Penguin Putnam; ISBN:0140282025

Strout, Joe; "Mind Uploading: an alternative path to immortality" in: *Cryonics* (1998, Vol. 19); pg. 26–30

Cryonics

Ettinger, R; *The prospect of immortality* (1965); http://www.cryonics.org/book1.html

Smith, George P; *Medical-Legal Aspects Of Cryonics : prospects for immortality* (1983); Port Washington Associated Faculty Press; ISBN:0867330503

Wowk, Brian; *Cryonics: Reaching for Tomorrow* (1991); Alcor Life Extension Foundation; ASIN:1880209004

Caloric Restriction

Heilbronn, LK & Ravussin, E; "Calorie restriction and aging: review of the literature and implications for studies in humans" in: *American Journal of Clinical Nutrition* (2003, Vol. 78); pg. 361–9

Lane, MA & Mattison, J & Ingram, DK & Roth, GS; "Caloric restriction and aging in primates: Relevance to humans and possible CR mimetics" in: *Microscopy Research and Technique* (2002, Vol. 59); pg. 335–8

Masoro, Edward J; *Caloric Restriction: A Key to Understanding and Modulating Aging* (2002); Elsevier Health Sciences; ISBN:0444511628

Weindruch, Richard, and Walford, Roy; *The Retardation of Aging and Disease by Dietary Restriction* (1988); Charles C. Thomas, Springfield, IL

Ethics & Philosophy: Basic

Brennan, H; *Death: the great mystery of life* (2002); Carroll & Graf Publishers ISBN: 0786712171

Broderick, Damien; *The Last Mortal Generation: How Science Will Alter Our Lives in the 21st Century* (2000); New Holland Publishers, Ltd.; ASIN:1864364408

Callahan D; *Setting Limits: Medical Goals in an Aging Society With "a Response to My Critics"* (1995); Georgetown University Press; ISBN:0878405720

Fukuyama, Francis; *Our posthuman future: Consequences of the Biotechnology Revolution* (2003); Picador, USA; ISBN:0312421710

Hardwig, John; "Is there a duty to die?" in: *Hastings Centre Report* (2000)

Kass LR; *Life, Liberty and the Defense of Dignity: The Challenge for Bioethics* (2002); Encounter Books; ISBN:1893554554

Kass Leon R; "L'Chaim and Its Limits: Why Not Immortality?" in: *Journal of Religion and Public Life* (2001, Vol. 113); pg.17–

Kaufman, Wallace; *No Turning Back: Dismantling the Fantasies of Environmental Thinking* (1995) Perseus Book Group; ASIN:0465051197

Lawson, Chris; *The Tithonus Option is Not an Option* (1999) presented at the SF Worldcon held in Melbourne, Australia

Perry, John; *A Dialogue on Personal Identity and Immortality* (1978); Hackett Publishing Company; ISBN:0915144530

Perry, Michael; *Forever For All: Moral Philosophy, Cryonics, and the Scientific Prospects for Immortality* (2000); Universal Publishers; ISBN:1581127243

Novels

Broderick, Damien; *Transcension* (2003); Tor Books; ISBN:0765303701

Dewdney, Christopher; *Last Flesh: Life in a Transhuman Era* (1998); HarperCollins Canada; ISBN:0006384722

Egan, Greg; *Permutation City* (1998); Gollancz; ISBN:0752816497

Halperin, James; *The First Immortal* (1998); Ballantine Books; ISBN:0345420926

Harrington, Alan; *The Immortalist* (1977) Ten Speed Pr.; ASIN:0890871353

Heinlein, Robert; *The Door Into Summer* (1986); Del Rey Books; ISBN:0345330129

Minsky, Marvin; Harrison, Harry; *The Turing Option* (1992); Warner Books; ISBN:0446364967

Nagata, Linda; *Tech Heaven* (1995); Bantam; ISBN:0553569260

Stephenson, Neil; *Snow Crash* (2003); Bantam; ISBN:0553380958

Vinge, Vernor; *A Deepness in the Sky* (2000); Tor Books; ISBN:0812536355

Wilson, Robert; *Prometheus Rising* (1992); New Falcon Publications; ISBN:1561840564

Further Resources

Tipler, Frank J; *The Physics of Immortality* (1995); Anchor; ISBN:0385467990

United States Congress, Senate. Special Committee on Aging: All hearings; For sale by the Supt. of Docs. U.S. G.P.O. Congressional Sales Office.

AUTHORS

William Sims Bainbridge, Ph.D.

http://mysite.verizon.net/william.bainbridge/
Deputy Director for the Division of Information and Intelligent Systems at the National Science Foundation, where programs supervised include *Nanoscale Science and Engineering*. Former director of several programs in that same foundation. Author of numerous books and articles on Technology, Sociology and religious studies. 1990-1992 Professor and Department Chair, Department of Sociology and Anthropology, Towson University (tenured). Ph.D. in Sociology (Harvard University).

Ben Best

http://www.benbest.com/
Senior Programmer/Analyst for a Canadian bank. He lives in Toronto, Canada and has served as Treasurer of Toronto MENSA and as President of the Toronto ACM APL (mathematical computer-language) Special Interest Group. Director of the Cryonics Society of Canada and CryoCare Officer. Ben is an essayist who has written on a wide variety of subjects.

Russell Blackford, Ph.D.

http://www.users.bigpond.com/russellblackford/
Writer, critic, and student of philosophy, based in Melbourne. First class honors degrees in both Arts and Law. Publications include *Strange Constellations: A History of Australian Science Fiction* (1999; co-written with Van Ikin and Sean McMullen), the science fiction trilogy *Terminator 2: The New John Connor Chronicles* (2002-2003), and many articles, reviews, and short stories. He is currently a graduate student and a sessional lecturer in the School of Philosophy and Bioethics, Monash University.

Nick Bostrom, Ph.D.

http://www.nickbostrom.com/
British Academy Research Fellow at Oxford University, Dr. Bostrom is co-founder of the World Transhumanist Association and has a background in physics, computational neuroscience, artificial intelligence, and philosophy.

Manfred Clynes, Ph.D.

http://www.microsoundmusic.com/clynes/
Professor at the Lombardi Cancer Center in Georgetown University in Washington, DC, Clynes grew up listening to recordings of Pablo Casals' transcendent classical performances, and often the music would transport him. " I used to have these wonderful moments of ecstasy that seemed tremendously important." he says, "And I assumed that other people had them all the time; only much later I found out that wasn't the case." During a presentation at a NASA conference in 1960, Professor Clynes coined the term *cyborg*, comb-

ing the two terms *cybernetic* and *organism*, as a concept for humans to survive space travel.

Robert A. Freitas Jr., J.D.

http://www.rfreitas.com/

Senior Research Fellow at the Institute for Molecular Manufacturing, a molecular nanotechnology think tank in Palo Alto, California, Freitas was the first to publish a detailed technical design study of a medical nanorobot in a peer-reviewed mainstream biomedical journal. Freitas is the author of Nanomedicine, the first book-length technical discussion of the medical applications of nanotechnology and medical nanorobotics.

Marc Geddes

http://www.prometheuscrack.com/

With a special interest in artificial intelligence and mathematics, Geddes is a freelance writer from Auckland, New Zealand. He has published a number of articles in various transhumanist media.

Aubrey de Grey, Ph.D.

http://www.gen.cam.ac.uk/sens/index.html

University of Cambridge, UK. Serves on the board of directors of the International Association of Biomedical Gerontology and the American Aging Association and on the editorial boards of the Journals *Rejuvenation Research*, *Mitochondrion* and *Antioxidants and Redox Signaling*. His research focus is to expedite the development of a true cure for human aging.

Raymond Kurzweil, Ph.D.

http://www.KurzweilAI.net

Raymond Kurzweil was the principal developer of the first omni-font optical character recognition technology, the first print-to-speech reading machine for the blind, the first CCD flat-bed scanner and the first commercially marketed large-vocabulary speech recognition software. Ray has successfully founded and developed nine businesses in OCR, music synthesis, speech recognition, reading technology, virtual reality, financial investment, medical simulation, and cybernetic art.

In 1999, Kurzweil received the US National Medal of Technology, the nation's highest honor in technology.

João Pedro de Magalhães, Ph.D.

http://senescence.info/

Microbiologist, research fellow in genetics at Harvard Medical School, Lipper Center for Computational Genetics, Boston, USA. His work on the biology of aging relates to cellular senescence, the telomeres, stress-response mechanisms, and Werner's syndrome. Dr. Magalhães develops computational approaches aimed at understanding aging from a genomic perspective.

Marvin Minsky, Ph.D.

http://web.media.mit.edu/~minsky/

Toshiba Professor of Media Arts and Sciences, and Professor of Electrical Engineering and Computer Science, at the Massachusetts Institute of Technology. His research has led to both theoretical and practical advances in artificial intelligence, cognitive psychology, neural networks, and the theory of Turing Machines and recursive functions. Professor Minsky was also one of the pioneers of intelligence-based mechanical robotics and telepresence. When a Junior Fellow at Harvard, he invented and built the first Confocal Scanning Microscope.

In 1959, Minsky and John McCarthy founded what became the MIT Artificial Intelligence Laboratory, and his long tenure as its co-director placed his imprint upon the entire field of Artificial Intelligence.

Brad Mellon, Ph.D.

Director of Chaplain Services, Frederick Mennonite Community, Frederick, PA, and Adjunct Professor, Bethel Seminary of the East, Dresher, PA (Greater Philadelphia area).

Born in a suburb of New York City in 1949, attended Houghton College, graduated with a degree in Sociology in 1967 and earned a Master of Divinity degree from Biblical Theological Seminary in 1980. Between 1985 and 1996, Dr. Mellon received the Master of Sacred Theology degree, became involved in postgraduate work at Dropsie University, and satisfied the requirements for the Doctor of Philosophy in Hermeneutics at Westminster Seminary.

Max More, Ph.D.

http://www.maxmore.com/

Dr. Max More is an internationally acclaimed strategic futurist who writes, speaks, and organizes events about the fundamental challenges of emerging technologies. Dr. More co-founded and is Chairman of Extropy Institute. As a leading transhumanist thinker, Max strongly challenges traditional, limiting beliefs about the possibilities of our future. Director of Content Solutions at ManyWorlds, Inc., his academic background includes a degree in Philosophy, Politics, and Economics from Oxford University, and a Ph.D. in Philosophy from the University of Southern California.

Eric S. Rabkin, Ph.D.

http://www-personal.umich.edu/~esrabkin/
Professor of English Language and Literature at the University of Michigan in Ann Arbor. As a teacher, Rabkin is especially known for his large, popular lecture courses on science fiction and fantasy, and for his many teaching innovations, including the development of the highly successful Practical English writing program.

Michael Rose, Ph.D.

http://ecoevo.bio.uci.edu/Faculty/Rose/Rose.html
Professor at the University of California, Irvine, focusing on the evolution of life histories and genetic systems, Michael has studied both the evolution of mortality late in life the evolution of fecundity late in life. Together with Dr. L.D. Mueller, he has developed evolutionary theories for late life plateaus, sometimes called immortality, and tested them using the fruit fly.

Mike Treder

http://www.incipientposthuman.com/
Executive director of the Center for Responsible Nanotechnology, is a business professional with a background in technology and communications company management. Serves on the Boards of Directors of the Human Futures Institute and the World Transhumanist Association, is a member of the Executive Advisory Team for the Extropy Institute, and developer of the *Incipient Posthuman* website.

Shannon Vyff

Mother of three, studying for a Master's in Nutritional Therapy, Shannon is a long-term leader for La Leche League International. She and her husband are members of the Caloric Restriction Society. If needed, she and her family have made arraignments to be cryonically frozen at the Alcor Life Extension Foundation.

Michael D. West, Ph.D.

http://www.michaelwest.org

Received his B.S. degree from Rensselaer Polytechnic Institute in 1976, his M.S. degree in biology from Andrews University, and his Ph.D. degree from Baylor College of Medicine in 1989 studying the molecular biology of cell senescence. Founder of Geron Corporation of Menlo Park, California and later Origen Therapeutics of South San Francisco, California. He is currently Chairman of the Board of Origen. He has extensive academic and business experience in age-related degenerative disease, telomerase molecular biology, and human embryonic stem cells.

Brian Wowk, Ph.D.

http://www.21cm.com/

Physicist, cryobiologist, and Senior Scientist at 21st Century Medicine, Inc., in Rancho Cucamonga, California, a company specializing in low temperature preservation of tissue and organs for medical applications. His current interests include the design of molecules that specifically bind and inhibit growing ice crystals ("ice blockers"), and the physics and engineering issues of banking transplantable organs. Dr. Wowk is one of the few scientists in the world that are directly

involved with preserving large organs at cryogenic temperatures. He has also been a close observer of the controversial
field of cryonics for two decades.

Publishers Notes

The Immortality Institute

The Institute represents the tip of the sword in the life extension movement by advancing its mission to "conquer the blight of involuntary death." With a number of projects and a growing worldwide membership base, the Institute has blossomed in scope and scale since its inception in 2002. As a non-profit educational organization, the Institutes' success rests fully on the backs of a highly motivated and totally volunteer support team.

http://ImmInst.org

Libros En Red

Online since May 2000 and (according to traffic-ranker Alexa) in the top position among e-book publishers and sellers in Spanish.

350,000 registered members receive LibrosEnRed's monthly newsletter. Books are published both in electronic formats (as e-books) and in paper (using a print-on-demand system). They are sold in our virtual bookstore, in 300 affiliate sites, and through Amazon.com and BN.com.

http://LibrosEnRed.com

THANKS

Our thanks go to all those who made this book possible. Many thanks first and foremost to the online community of the Immortality Institute for encouragement, critique and many valuable suggestions. Special thanks to Michael Anissimov, Justin Corwin, Kevin Perrott, Kenneth X. Sills, Don Spanton and Alexandra Stolzing for their invaluable help and advice in the editorial process. Thanks to Gustavo Faigenbaum, Vanesa L. Rivera and all the patient and professional staff at Libros en Red. Warm thanks to the authors who have agreed to forego all potential revenue from this publication for the benefit of the Institute. And finally, thanks to those scientist and visionaries who are working on the scientific conquest of death this very moment.

Without them this book and the ideas discussed herein would never have been conceivable.